YALE UNIVERSITY

MRS. HEPSA ELY SILLIMAN

MEMORIAL LECTURES

Paleontology and Modern Biology

PALEONTOLOGY

and Modern Biology

by David Meredith Seares Watson

Jodrell Professor of Zoology
and Comparative Anatomy, University College
University of London

NEW HAVEN : YALE UNIVERSITY PRESS
London · Geoffrey Cumberlege · Oxford University Press

THE SILLIMAN FOUNDATION

IN THE YEAR 1883 a legacy of eighty thousand dollars was left to the President and Fellows of Yale College in the city of New Haven, to be held in trust, as a gift from her children, in memory of their beloved and honored mother, Mrs. Hepsa Ely Silliman.

On this foundation Yale College was requested and directed to establish an annual course of lectures designed to illustrate the presence and providence, the wisdom and goodness of God, as manifested in the natural and moral world. These were to be designated as the Mrs. Hepsa Ely Silliman Memorial Lectures. It was the belief of the testator that any orderly presentation of the facts of nature or history contributed to the end of this foundation more effectively than any attempt to emphasize the elements of doctrine or of creed; and he therefore provided that lectures on dogmatic or polemical theology should be excluded from the scope of this foundation, and that the subjects should be selected rather from the domains of natural science and history, giving special prominence to astronomy, chemistry, geology, and anatomy.

It was further directed that each annual course should be made the basis of a volume to form part of a series constituting a memorial to Mrs. Silliman. The memorial fund came into the possession of the Corporation of Yale University in the year 1901; and the present work constitutes the thirty-first volume published on this foundation.

v

FOREWORD

WHEN I received an invitation to deliver the Silliman Memorial Lectures for 1937, I appreciated the honor of contributing to that famous series and felt that it was an occasion on which to offer apologia pro vita mea.

Fossil vertebrates present a field of scientific work of a special kind, which needs for its exploitation a familiarity with several branches of knowledge but which gives promise of contributing to our understanding of the mechanism of evolution by allowing us to study the actual history of the development of some animals. The reliability of such evolutionary series rests entirely on the validity of the methods of reasoning by which they have been determined, and I know of no discussion of the intellectual respectability of the processes of paleontological work.

I have therefore devoted this series of lectures to a consideration of certain paleontological problems, on which I have myself worked, placing them in relation to other more general considerations and trying to make clear the assumptions on which their discussion really rests. And I have tried to show that the fundamental assumptions may properly be made.

The threat of war on my return to England in 1937 made my work on the Agricultural Research Council more onerous, and the outbreak of war in 1939 led to the whole of my energies being devoted to matters concerned with the food policy of Great Britain. This must be my excuse for the very long delay which has occurred between the delivery of these lectures and their publication.

Much work in the fields with which I deal has been published in the past twelve years, increasing our knowledge of fact and throwing new light on the problems I discussed. But I find little of importance to alter, though much to add, and the need for some discussion of the methods of paleontology remains as great as it ever was. These lectures are therefore published now essentially as they were delivered, but with the addition of a series of appendices which relate the impact of more modern discoveries on the subject of each lecture.

CONTENTS

Foreword   vii

I.   Verified Predictions   1

II.   Uniform Change   31

III.   The Meaning of Structural Change to the Animal   53

IV.   Origins   79

V.   Early Reptile History   105

VI.   The Approach to a Mammal Structure   136

VII.   Adaptive Change   165

VIII.   Possible Machinery   182

Bibliography   204

Index   213

ILLUSTRATIONS

1. Head of a dogfish — 5
2. Head skeleton of dogfish — 5
3. Head nerves of dogfish — 7
4. Head of embryo dogfish — 8
5. Head of *Climatius* — 9
6. Head of *Euthacanthus* — 9
7. Head skeleton: *Acanthodes* — 10
8. Head skeleton: *Macropetalichthys* — 11
9. Diagram of head of *Cephalaspis* — 13
10. Labyrinthodont ornament — 34
11. Early amphibian skulls — 35
12. Types of amphibian vertebrae — 37
13. Capitosaurine skull roofs — 40
14. Capitosaurine palates — 42
15. Skull structure of a labyrinthodont — 43
16. Capitosaurine occiputs — 48
17. Brachyopid skulls — 53
18. Brachyopid palates — 54
19. Brachyopid occiputs — 56
20. Fish-eating labyrinthodonts — 57
21. Palates of fish-eating labyrinthodonts — 58
22. Capitosaurine bases of skulls — 61
23. Trematosaur occiputs — 63
24. *Eryops* skeleton — 65
25. *Cyclotosaurus* skeleton — 65
26. Head movement: *Eryops* — 67
27. Head movement: *Cyclotosaurus* — 68
28. Lower jaws: *Eryops* and *Capitosaurus* — 69
29. Capitosaur occiputs — 71
30. Capitosaurine brain cases — 75
31. Fish to amphibian palate — 83
32. Primitive labyrinthodont locomotion — 86
33. Labyrinthodont palates — 88
34. Lepospondyl skulls — 90
35. Lepospondyl palates — 91
36. Fish to amphibian skull: top — 92

37. Fish to amphibian skull: side   93
38. Fish to amphibian skull: bottom   96
39. Axolotl   98
40. Gills of *Branchiosaurus*   100
41. Neotenous labyrinthodont from below   102
42. Neotenous labyrinthodont from the side   102
43. Swimming and terrestrial labyrinthodonts   107
44. *Seymouria* and a Carboniferous labyrinthodont: skulls   108
45. *Seymouria* and a Carboniferous labyrinthodont: palate and occiput   109
46. *Seymouria* and a Carboniferous labyrinthodont: vertebrae   111
47. Atlas and axis of *Seymouria*   113
48. *Seymouria* and two primitive reptile skulls   117
49. Primitive pelycosaur skull   119
50. Pelycosaur skull shapes   120
51. Pelycosaur skeleton   120
52. Gorgonopsid skeleton   122
53. Skull of a cynodont   124
54. Ear ossicles   137
55. Skulls of mammal and lizard   140
56. Ear of lizard   141
57. Ear of mammal   142
58, 59, 60. Palate and jaw from amphibia to mammal   145–147
61, 62, 63, 64, 65. Skull of gorgonopsid: side, top, palate, occiput, lower
      jaw   149–151
66. Diademodon: lower jaw   153
67. Young opossum, ear ossicles   157
68. Skull, lower jaw, and ear of cynodont   158
69. Skeleton of a plesiosaur   166
70. Plesiosaur paddles   168
71. Plesiosaur shoulder girdle   171
72. Mechanics of plesiosaur shoulder girdle   173
73. Series of elasmosaur shoulder girdles   174
74. Body form of elasmosaurs   175
75. Body form of long-headed plesiosaurs   176
76. Shoulder girdle of large-headed plesiosaurs   177
77. Restorations of ichthyosaurs   180

I. VERIFIED PREDICTIONS

Fossils, the preserved remains of animals and plants dug out of the earth, have long attracted the attention of men. In a neolithic grave in England a man found buried was surrounded by a ring of fossil sea urchins preserved as flints from the chalk, carefully selected, no doubt, because of their uniform shape and attractive ornamentation. The classical Greeks believed the Island of Samos was a former home of giants because they found near Mitylene a bed full of large bones, now known to be those of rhinoceros and other great beasts. But the first considerable advance in our knowledge of fossil vertebrates was made by the systematic studies of Cuvier. He and his successors compared fossil bones and teeth one by one with those of still living animals. Their work was carried out at a time when it was only gradually becoming clear that all fossils were not of the same age, that in fact they were not exclusively the remains of animals drowned in Noah's flood.

After the discovery by William Smith that "strata could be identified by the organized fossils they contain," the science of paleontology assumed a new importance because it was seen to be a necessary aid in stratigraphy; and at the same time paleontology itself gained that feature which really characterizes it. It became a historical science, its subject the history of life in the world. As such it has its own niche; it is distinguished from zoology by the existence of this time element and from geology by the fact that fossils were once parts of living beings only to be understood by zoologists or botanists.

When, in consequence of the publication of Darwin's *Origin of Species* in 1859, the theory of evolution came to be discussed and finally accepted by all biologists, paleontology gained a new value. It was clear from the first that the evidence of fossils was vital for all discussions of the general theory. Indeed, so convinced was Darwin that the evolution theory would stand or fall by the nature of paleontological record that he devoted two chapters of his book to the discussion of this matter. How well based these were and how great was the foresight they displayed has since become evident. The general agreement between the actual facts of the seriation of fossil animals and plants and the expectation on any evolutionary theory is so perfect that in itself it affords the strongest proof of Darwin's thesis.

1

But we can now go much further. The discoveries made in the western United States by Othniel Charles Marsh and those who followed him soon enabled the story of the evolution of the horse to be traced, first in outline, but soon with a mass of corroborative detail. And the story of the horse has been followed by those of the camel, deer, pig, dog, elephant, and many others, until the evolutionary history of the mammals has become known to us in detail almost excessive in its amount.

But in part because its very abundance makes it difficult to understand, the evidence afforded by these well-founded phylogenies has seldom been systematically used in an attempt to elucidate the actual mechanism of evolution. Indeed paleontologists on the whole have not paid the attention it deserves to our rapidly growing knowledge of genetics and of developmental physiology.

It is my intention in this book to consider how far it is possible to establish general principles which apply to all kinds of fossil materials and which arise naturally from their study, and then to attempt to relate them to other recent biological conceptions.

During the past thirty years I have worked on the lower vertebrates, on fish, amphibia, and reptiles, and have thus, unlike the student of fossil mammals and of the majority of invertebrates, been driven into the consideration of major evolutionary steps: the transition from water-breathing fish to air-breathing amphibia; from amphibia which passed their early stages in water to reptiles which, because their eggs included a store of the small quantity of water necessary to allow of full development, could colonize deserts; and finally the gradual transformation of a cold-blooded reptile to a warm-blooded mammal.

My work has lain in a field wide in its scope but one in which the available stages are much further removed from one another than those at the disposal of mammal paleontologists.

It seems to me essential to understand the real nature of the intellectual processes we use and to establish their validity, and it is clear that the most striking confirmation of the logic of an argument is the verification of predictions. It is my present purpose to show how some such predictions have been made and how the proper description of fossils has demonstrated them to be correct.

The object of every paleontologist is to make genealogical trees, to bring together animals of successive periods which stand to one another in the relation of father to child. This involves two quite distinct processes: the relationship must be established and the order of appearance of the fossil forms must be accurately known. Neither can ever be a matter of direct observation; we cannot breed our subjects as the geneticists do, and we are not spectators of the process to whom the facts of succession are obvious!

The only evidence we can have of relationship is derived from similarity of structure. In the case of close series this similarity may be evident to a layman, who if quick-witted may well, as the Chinese peasants do, recognize the skull of the three-toed horse *Hipparion* as that of a donkey. But this intelligent layman will inevitably be mislead by all such cases of resemblance as those between edible and distasteful insects, which are called mimicry. The existence of real, fundamental structural resemblance can only be established by using the intellectual processes which characterize that branch of zoology called morphology. Morphology is a form of logical thought remarkable in that it is not mathematical; indeed, its essential elements, being, as they are, qualities, are not susceptible of mathematical expression. I do not wish to imply that mathematical treatment may not be a valuable, indeed an essential, tool in morphology, but it is always ancillary. It can be used to clear inessential from essential differences, but the fundamental facts will remain free from symbolic expression. The basis of morphology lies in that distinction, first pointed out by Richard Owen, between homology and analogy. In his illustration the forelimb of man and the wing of a bird are homologous organs, while the wings of birds and insects are analogous structures. Homologous structures need not, although they may, serve the same functions, while functional similarity is the whole basis of the comparison between analogous parts.

Thus the evidence on which two organs are held to be homologous must be anatomical, all differences being of such a nature that they can be shown to meet functional needs. Therefore in practice it is necessary to investigate the functional significance of each structure considered, to find out how it works and how its actual character is related to this usage, and by stripping off these adaptive devices to arrive at the essence of the organ, considered merely as a part of the make-up of an animal. When this has been done for the structures in two animals under consideration it will be found that if they are homologous the residue, the essence, of each will be the same. In terms of the evolution theory homologous structures are those which have been derived from the same parts of a common ancestor. It should therefore be possible to trace in fossil material the steps by which changes of function, necessarily involving modification of structure, actually take place and to determine how it is that the intermediate stages of such transformations, which are sometimes of a grotesque kind, remain parts of a successful animal.

The conception of homology has been widened to cover the resemblance between parts of the successive segments of the same individual animal, when, as in the earthworm and in ourselves, the body becomes transversely divided into short lengths, each theoretically identical with that before and that behind itself.

The criteria used in establishing both types of homology are the same. It is difficult to draw up in set and general terms the nature of these criteria, but it is easy by the consideration of actual cases to illustrate their general application. In suitable cases it is possible to go as far as the prediction of a hypothetical ancestral stage and then to find the fossil form in which the predicted structure can actually be seen.

One of the most enthralling of all morphological studies has been the investigation of the segmentation of the head of vertebrates. Begun by Oken and Owen more than a hundred years ago, this investigation was placed on a sound basis; it was, in essence, solved by the embryological studies of Francis Maitland Balfour in 1876. Subsequent work has confirmed and very much extended his conclusions.

If we examine the head of any shark or dogfish we find on the flank of the fish, between the angle of the mouth and the pectoral fin, a series usually of five vertical slits, decreasing in size from front to back but otherwise similar. Each has a small flap covering it and leads through an enlarged chamber into the alimentary canal (Fig. 1). In a live fish it is found that a stream of water is regularly but intermittently expelled from each of these slits. It is known that the whole process is part of the mechanism of respiration, and that oxygen is actually taken into the fish's body through the thin skin which covers bunches of delicate filaments attached to the anterior and posterior surfaces of the enlarged gill spaces. In some dogfish and sharks there is a hole, much smaller than an ordinary gill slit, lying high up just behind the eye, which during life serves for the intake of water. Let us consider the nature of this hole, which is called the spiracle. It is clearly functionally unlike the gill slits because water goes through it in the opposite direction. But it is structurally similar because it connects the alimentary canal with the skin, and its internal opening lies in line with those of the gill

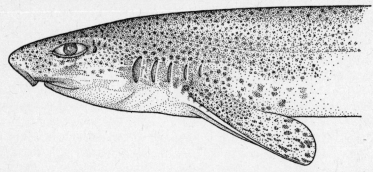

FIGURE 1. Head of the common dogfish, *Scyllium canicula.* The row of five gill slits can be seen, their depth decreasing from front to back. The spiracle is a small black hole immediately below and behind the posterior corner of the eye.

slits. Furthermore it retains a trace of gill filaments on its anterior wall. It is thus possible that the spiracle is a serial homologue of the gill slits, and it should be possible to show that its relation to all surrounding structures is the same in essence as those between gill slits and their surroundings.

The easiest structure to investigate is the skeleton. In Fig. 2 it

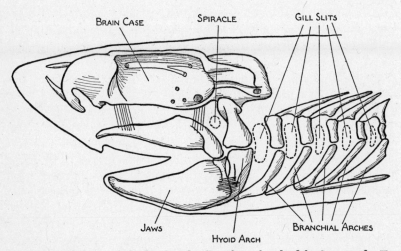

FIGURE 2. Skeleton of the head and gill arches of a dogfish, *S. canicula.* To show the branchial arch skeleton and the relation of the spiracle to the hyoid arch and the jaws. Labial and extrabranchial cartilages omitted.

is clear that each gill slit is separated from its neighbor by a hoop made up of a series of four cartilaginous rods on each side, movable though attached to one another. The first gill slit passes between the first of these branchial arches and the part of the skeleton (the hyoid arch) which lies between it and the spiracle. The skeleton of the hyoid arch resembles that of an ordinary branchial arch in that it is a jointed hoop of rods of cartilage, the most important joint being in the same place, but differs in containing only two rods instead of four. The spiracle in turn is bounded in front by two cartilages, the jaws, which articulate in the same line as the hyoid arch. Further-more, unlike all the branchial arches, the hyoid arch and the jaws are tied tightly together by a short, tough ligament below the spiracle, and the jaws are attached to the brain case mainly through the up-per element of the hyoid arch.

The resemblance between the hyoid arch, the mandibular arch, and the typical branchial arches is sufficient to suggest their homology but clearly not sufficiently close to establish it. We are thus driven to the consideration of other associated structures. Many may be used: the muscles or the blood vessels are capable of giving good evidence in favor of the serial homology of these structures; but for certain reasons the nerves supplying the gill arches are the most convenient.

Each of the gill pouches is supplied by a single nerve which forks over its upper edge and has in addition a small branch passing in-ward and forward over the main food passage of the alimentary canal. This nerve arises from a ganglion, which is itself connected with the hindbrain at a definite level. The three branches of this nerve serve taste and other senses, but only that which lies behind the gill slit goes to muscles and transmits those nerve impulses which, by causing contractions of these muscles, bring about movements of the branchial arches. Such a nerve is called branchiomeric (Fig. 3). The branchiomeric nerves which supply the last four gill slits are so tightly packed together at their origin from the brain that they form essentially one joint nerve (the tenth or vagus). The nerve to the first gill pouch is completely independent and can be directly com-pared with the facial nerve to the spiracle. It is at once evident that the two agree exactly; they branch in the same way, the branches have the same relation to the actual perforation, and they have the same function.

When however we pass to the trigeminal nerve supplying the

corner of the mouth we find a less exact correspondence. The nerve
branches to pass above (that is, in front of) the mouth and below
and behind it, but there is no branch passing inward over the roof
of the mouth. Both branches are sensory, though the senses they
subserve are touch and the ordinary senses of the skin, not taste.
But, as in a branchiomeric nerve proper, only the posterior branch
has motor fibers to the muscles which move the jaws. The significance of these differences can readily be understood if the functions
they carry out are considered; and if we take for our comparison
a lamprey instead of a shark we find a trigeminal nerve identical
in its composition to the facial nerve.

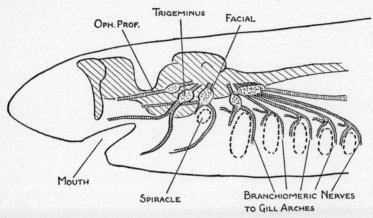

FIGURE 3. The branchiomeric nerves of the dogfish, *S. canicula.* To show
that the facial nerve which supplies the spiracle is related to that opening
exactly as the other branchiomeric nerves are to the gill slits. In this
drawing the acousticolateral system of nerves and the somatic system of
nerves are omitted. The obliquely lined structure is the brain; ganglia are
dotted.

Thus consideration of their nerve supply supports the view that
the jaws, the hyoid arch, and the branchial arches are homologous
structures, and that the spiracle is homologous to a gill slit. Further
confirmation is afforded by an examination of these structures in
the embryo dogfish (Fig. 4), and from such an investigation we gain
support for the view that the small size and different function of the
spiracle are secondary features resulting from the use of the hyoid
arch to support the jaws and to connect them to the brain case and
hence to the body as a whole.

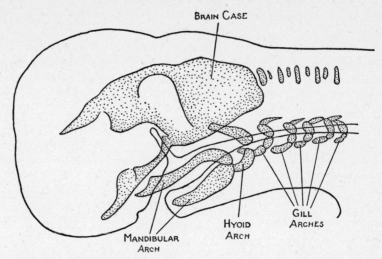

FIGURE 4. The developing head skeleton of a dogfish, *S. canicula*. To show the general similarity of the mandibular arch (jaws), hyoid arch, and gill arches. Redrawn from Goodrich, 1918.

From this discussion it is possible to proceed to the prediction that there must once have existed fish in which the hyoid arch and the jaws consisted of four separate cartilages on each side, as did the branchial arches, and in which the hyoid arch was separated from the jaws by a full-sized gill slit and did not in any way contribute to their support. As such animals would be fundamentally more primitive than true fishes they might be expected to precede them in time and to exhibit a great variety of adaptations.

Such creatures are known, the central group being the Acanthodii. In the most primitive of them, *Climatius* (Figs. 5 and 6), there is a series of very long gill slits, each overhung by an opercular fold, and the largest of these slits is the homologue of the spiracle, the operculum covering it being carried on the jaws. The cartilages themselves are scarcely known in *Climatius,* but they are completely known in the later *Acanthodes* (Fig. 7). In this fish the mandibular arch is composed of four primary elements, and the hyoid perhaps of four, certainly of three; both are clearly homologous with gill arches. The spiracle is an immense gill slit, guarded throughout its extent by gill rakers so set as to form a food-collecting sieve.

Thus the original prophecy, made implicitly at least fifty years ago, is fulfilled and its extensions are justified. The time at which the acanthodians were most abundant and varied, the Lower Old Red

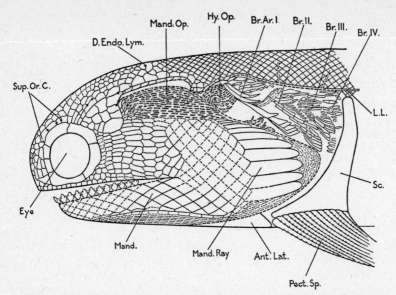

FIGURE 5. Head of the acanthodian fish *Climatius reticulatus* from the Lower Old Red Sandstone of Forfarshire. To show the series of opercula on the mandibular, hyoid, and branchial arches, which indicate that the spiracle was a full-sized functional gill slit. *Mand. Op.* and *Mand. Ray,* the upper and middle parts of the mandibular operculum. *Hy. Op.,* hyoid operculum, guarding the first gill slit. *Br. I–IV,* the opercular bones of the four branchial arches. From Watson, 1937.

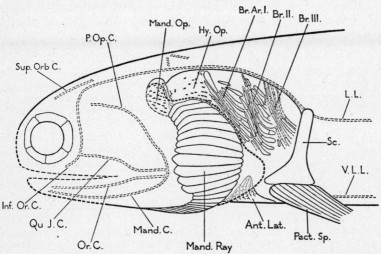

FIGURE 6. Head of the acanthodian fish *Euthacanthus macnicoli* from the Lower Old Red Sandstone of Forfarshire. Compare with Fig. 5, which in principle it greatly resembles. The reference letters have the same significance. From Watson, 1937.

9

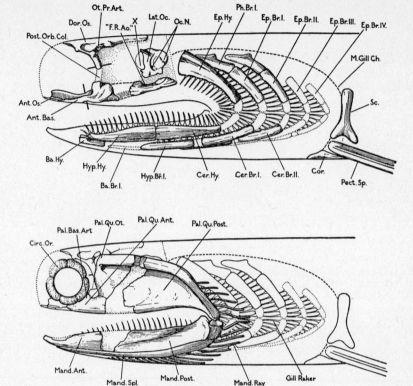

FIGURE 7. Head skeleton of *Acanthodes,* from the Lower Permian of Lebach shales of Saarbruck. This figure is intended to show that the long continuous series of gill rakers projecting forward from the hyoid arch imply that the hyoid cleft was a full-sized gill slit, not a spiracle.

In the lower figure the mandibular arch skeleton is in position and the attachment of the mandibular operculum to it is clearly shown. In the upper figure the mandibular arch has been removed so as to expose the whole extent of the hyoid arch skeleton. The relevant reference letters are: *Ba. Hy.,* basihyal; *Cer. Hy.,* ceratohyal; *Ep. Hy.,* epihyal; *Hyp. Hy.,* hypohyal; *Mand. Ant.,* anterior bone in the lower jaw; *Mand. Ray,* a ray in the mandibular operculum; *Mand. Spl.,* the mandibular splint, a membrane bone; *Pal. Qu. Ant.,* the anterior bone in the upper jaw; *Pal. Qu. Ot.,* otic process of the upper jaw; *Pal. Qu. Post.,* the posterior bone in the upper jaw. From Watson, 1937.

Sandstone, antedates the appearance of true fish, and the great group, the Aphetohyoidea, which may be set up for their reception, must include also many of the most bizarre of the Paleozoic vertebrates, the Arthrodira and their allies.

The analysis of the segmentation of the head of vertebrates can be carried much further and predictions made as a result of an immensely long and difficult argument. The work of Professor E. A. Stensiö of Stockholm enables us to confirm these predictions.

If we go on from the point reached in the preceding argument, that the jaws are homologous with a branchial arch, we can suggest that the corner of the mouth, which is related to its trigeminal nerve

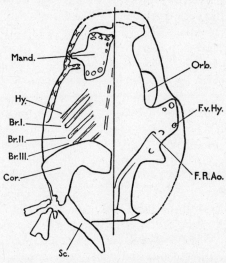

FIGURE 8. The head of *Macropetalichthys,* to show that in this animal the hyoid arch, *Hy.,* has no attachment to the lower jaw, *Mand.,* and that the creature may be placed in the same class, the Aphetohyoidea, as the acanthodians. From Watson, 1937, after Broili.

as a gill slit is to its own branchiomeric nerve, is actually a gill slit. The argument now necessarily turns to a new set of structures and to the evidence presented by early stages of development. In all vertebrates the first appearance of segmentation is the breaking up by slits of a formerly continuous strip of tissue (the paraxial mesoderm) into short blocks, each of which among other things gives origin to segmentally arranged trunk muscles. These arose in the first instance to facilitate the characteristic undulatory swimming movements of vertebrates. The presence of these segmental muscles imposes a segmental structure on the nerves which supply them, and on the gill slits which can only pass out between them.

In such fish as the sharks the segmental blocks, the somites, which form muscle continue forward into the head, and each has its own nerve. Here as in all vertebrates the three anterior pairs of somites form only the muscles which move the eyeballs. By a long process of reasoning it can be shown that the third somite is related to the facial nerve and the spiracle, the second to the trigeminal nerve and the corner of the mouth. The first somite is associated with a nerve called the ophthalmicus profundus which is not obviously comparable to a branchiomeric nerve and is not in any way related to any existing structure comparable to a gill slit. It is possible to interpret the whole condition as a relic of the structure of an extremely primitive vertebrate in which the mouth was a very small hole with no jaws, behind which lay two pairs of gill slits, the first related to the profundus nerve and the second to the trigeminus. At this stage the vertebrates fed on minute food particles suspended in water or mixed in mud. When they changed their habits and began to eat large prey the mouth clearly had to be enlarged, and this enlargement, made by carrying the corners of the mouth backward, crushed out of existence the first pair of gill slits, the process being arrested when the mouth had absorbed the second pair. The mouth thus came into relationship with the branchial arch, which lay behind the second pair of gill slits, and the skeleton and muscles of this arch became the jaws and their musculature. That part of the head, including the mouth, which lay in front of the first gill slits was a remnant left behind by the process of cutting off somites and is represented in shark embryos by the anterior head cavity discovered by the American zoologist Miss Platt.

This hypothetical creature is completely realized in the long-known fossil, *Cephalaspis* (Fig. 9), in which Dr. Stensiö has shown that the ophthalmicus profundus and the trigeminus were typical branchiomeric nerves, each related in the ordinary way to a functional gill pouch.

Thus even the most extended flights of morphological argument have led to predictions which, within the past few years and very unexpectedly, have been verified by the description of actual fossil animals whose structure conforms exactly to expectation. The fact that the methods of morphology allow us to conduct coherent and self-consistent arguments leading to conclusions that have ultimately been verified by the discovery of fossil animals, enables us to use these methods in sorting out from the innumerable known fossils groups of animals with such similarities of structure that we may

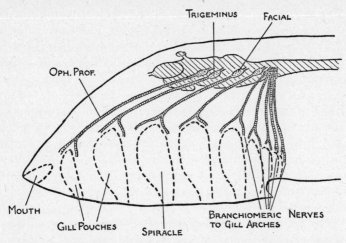

FIGURE 9. The head of the cephalaspid *Kiaeraspis*, to show the brain and the series of branchiomeric nerves and their relation to the gill pouches. The ophthalmicus profundus nerve is related to a gill pouch which has disappeared in all other vertebrates. The trigeminus nerve bifurcates around a gill pouch which occupies the position of the corner of the mouth of the gnathostome (jaw-bearing) vertebrates. An original interpretation founded on the figures and descriptions of E. A. Stensiö.

reasonably regard them as blood relations. It gives an intellectual respectability to our procedure.

APPENDIX

The older literature on the segmentation of the head in craniate vertebrates is very familiar and is most admirably summarized by Goodrich in *Structure and Development of Vertebrates,* chap. v. There have, however, been additional researches of great importance which have added to our conception of the process of cephalization at the same time that they have confirmed, by new evidence, the classical interpretation of the process.

The first work of importance on this matter was that by F. M. Balfour in 1877, who established beyond dispute the existence of a segmentation extending far forward to the tip of the notochord at any rate, and determined the existence of at least eight segments in the skull of an elasmobranch fish. The next work of importance, by van Wijhe, went beyond Balfour in showing that the muscles which move the eyeball are somatic muscles, homologous with those

of the trunk; that the six muscles come from three somites; and that
these somites correspond with the seventh cranial nerve for the most
posterior of them and with the fifth cranial nerve for the two anterior.
The next contribution came from the investigation, first of all by
Strong and then particularly by C. J. Herrick, of the composition
of the cranial nerves of all craniates, which showed the complete
similarity of the branchiomeric nerves and led to the recognition of
the fact that the fifth cranial nerve was really double, consisting of
the obviously branchiomeric trigeminal nerve, with its maxillary
and mandibular branches, while in the majority of craniates the first
branch of the fifth nerve, the ophthalmicus profundus, is really independent, the nerve being related to a segment in advance of that
supplied by the trigeminus.

The homology of the spiracle with the gill slits has never been dis-
puted. Whether the mouth is merely an enlarged gill slit or whether
the gill slit whose place it takes has been crushed out of existence by
a movement backward of the corner of the mouth is uncertain; nor
is it, in fact, a matter of importance. The establishment of the condition of the hyoidean cleft—the spiracle of fishes—in acanthodians
was difficult, but the evidence seems to me to be conclusive. It is as
follows: although the genus *Climatius* is represented by many individual specimens, only one was so preserved as to show the structure of the head with complete clearness. In this specimen (Watson,
1937, Pl. V), the fish was completely flattened at burial, crushed
obliquely so that the cheek and the dorsal surface of the head are
brought into one plane. The rock in which it was contained was
split along a plane which passes through the middle of the body,
the bones being seen chiefly from their inner surfaces, the top of
the head and right side of the face being preserved in one slab,
the intergular space and the left cheek on the other. The preservation is exquisite, even the most minute bones, perhaps only one tenth
of a millimeter across, being perfectly distinct. In this specimen the
gill region is perfectly shown, but the only hard parts belonging to
it are those of the dermal bone which overlaid it on its outer side.
The actual cartilages of the branchial arches are completely unossified. Posteriorly there are four series of small overlapping dermal
bones, of which the most obvious run obliquely downward and backward from the dorsal side of the body. From these extend backward
very small elements, lying in the skin, which together cover quite
completely the posterior part of a gill chamber whose limits are
clearly shown by the abrupt ending of the body scaling. Detailed

examination makes it obvious that each of these four represents the outer surface of a gill arch with a flap of skin projected backward until it rests upon the arch behind. Farther forward there is a small patch of scales which very evidently represents a similar opercular flap, and this patch is associated with a narrow downturning of the scaly cover of the upper part of the head, so that it is evidently the remains of the hyoid arch, and the known position of the ear in the fish makes its homology certain. But proceeding backward from a similar yet much larger downturning which lies behind the eye there are first of all the bony cheek and lower jaw; dorsally and posteriorly to those a very large operculum overrides the lower parts of the hyoid and all the branchial arches. This structure, first clearly seen in a single specimen, can be observed in several others in precisely the same form, and there can be no doubt as to the reality of the condition which I have described above.

The argument from these conditions is therefore this. It is evident that the first normal gill slit which lies behind the hyoid arch is covered by a valve which is the hyoid operculum, and similarly for all the other gill slits. (The conditions are in fact mechanically the same as in the dogfish, where each dorsoventrally short gill slit has its own opercular fold, a structure which is necessary in order to insure that the respiratory stream of water passes through the pharynx in one direction only.) The great operculum which is attached to the mandibular arches is extended forward far in advance of the hyoid arch in a way which makes it evident that the whole structure is indeed of mandibular origin; and it seems to me highly improbable that so complete an operculum should exist had the prehyoidean gill slit been reduced to a small, dorsally placed spiracle, as it is in elasmobranchs and in all bony fishes. Had such a reduction taken place the mandibular operculum would be no more than a small valve controlling the spiracular cleft, and the main operculum would, as in bony fishes, have been a development of the hyoidean operculum, which in fact it overrode. I therefore concluded that there was a prehyoidean gill cleft of full dorsoventral extent. After this structure had been determined I passed on to the consideration of other fish and found, in perfectly preserved material of the allied fish *Euthacanthus,* an exactly similar structure. Here however the hyoidean gill septum bears a series of small, elongated bones corresponding exactly to those which cover (in *Climatius* and *Euthacanthus*) the outer surface of all the branchial arches. This series of bones in the hyoid arch, which supports a small hyoidean oper

culum, passes down below the uppermost of the enlarged rays which
lie in the mandibular operculum. The elements of the two arches
meet at right angles so that it is obvious that one lay external to the
other. There is here then a clear indication of the continuance of
the hyoid arch mesial to the mandibular opercular fold, and the
same argument for the existence of a full-sized gill slit applies.

A small fish, *Brachiacanthus*, repeats the same story, and in the
genus *Mesacanthus*, of much more advanced general structure
though still an acanthodian, in a specimen which was so young that
no ossification had taken place in the cartilaginous jaws, the hyoid
arch can be seen with a coating of membrane bones along its length,
which extends so far ventrally that it terminates at the level of the
quadrate-articular joint and lies in front of that structure. In other
words there is a hyoid arch with a free outer surface coated with
dermal bone lying mesial to the inner surface of the palatoquadrate
cartilage in such a way that it is obvious that the two elements were
not in contact; furthermore the lower part of this hyoid arch supports a small series of dermal rays which can only have formed part
of the support of an operculum, that which covered the first gill slit.
But in this animal there is a very extensive mandibular operculum
extending backward so as to cover not only the prehyoid gill slit
whose existence is here so clearly shown but also the lower parts of
all the other posthyoidean gill slits. There is thus clear and direct
evidence of the existence in these fishes of a widely open prehyoidean
gill cleft, extending at least as far downward as the level of the jaw
articulation.

If now we pass to the one acanthodian in which the cartilages of
all the visceral arches are well ossified and completely shown as
solid objects—the genus *Acanthodes*—we find that the hyoidean
arch is visibly complete, shows no signs whatsoever of contact with
the palatoquadrate cartilage, and is indeed extremely similar in its
general structure to the succeeding branchial arches. Each branchial
arch of *Acanthodes* has attached to its anterior surface a continuous
series of gill rakers of remarkable and characteristic structure. These
are so arranged as to form a sieve; each is a flat sheet of calcified material, with its broad faces horizontal, and extends so far forward that
its tip rests upon the inner surface of the branchial arch in front. In
the case of the first branchial arch this series of gill rakers extends
from the pharyngobranchial to a point very near the anterior end of
the hypobranchial, an indication of the extreme length of the gill
slit which lay between the hyoid arch and that behind it. But the

hyoid arch, from the dorsal end of the epihyal to the extreme anterior end of the hypohyal, bears an identical series of gill rakers, and as from the structure of all the branchial arches it is evident that these structures crossed the gill slit in front of the arch to which they are attached, it seems evident that the exactly similar series in the hyoid arch must imply the presence of a full-length gill slit between the hyoid and mandibular arches. The conditions in *Acanthodes* thus confirm, by an entirely independent and different chain of reasoning, the conclusions derived from the conditions in the more primitive acanthodians.

It is necessary now to consider the actual cartilages which lie within the gill septa and form the jaws in acanthodians. These are known, with any completeness, only in *Acanthodes*, where the first and second gill arches can be seen each to contain a series of four bones movably articulated with one another in a chain running along the length of the arch. These are the pharyngobranchial, epibranchial, ceratobranchial, and the hypobranchial, which articulates with a median basibranchial element. The hyoid arch contains a similar chain extending from an epihyal downward and then forward to end in an articulation with a basihyal; and various authors—Reis, Jaekel, and Bashford Dean—have found a more dorsal element corresponding to the pharyngobranchial in the posterior arches. This element I have never seen, but I think it probable that it did really exist. But *Acanthodes* shows the existence, in the palatoquadrate cartilage and in Meckel's cartilage, of a series of four ossified elements which presumably represent independent cartilages. Two of these, the anterior and posterior palatoquadrate elements, lie in the upper jaw, the other two in the mandible. They correspond very precisely with the four bones in the branchial arches, and the main point of flexure (the articulation of the upper and lower jaws) lies in the same horizontal plane as the corresponding part in the hyoid and all other branchial arches. There seems no doubt that the same condition occurred in *Mesacanthus*. But in *Acanthodes* there is a new formation, not existing in the more primitive acanthodians, in the form of a third upper jawbone lying in the otic process—a structure which, as it is not represented in *Climatius*, for instance, seems to have been a neomorph. This articulates with the brain case lateral to the ear and immediately behind the eye. Such a process is obviously useful functionally in securing a firmer attachment of the jaws to the neural cranium and through it to the trunk. The upper jaw is considerably shorter than the lower jaw, as the upper part of each branchial arch is shorter than its lower end. It is therefore in many ways reasonable to assume that the jaw cartilages are homologous with those in the hyoid and branchial arches.

The evidence as to the existence of a full-sized prehyoidean gill slit in the other groups, of very different appearance from acanthodians, which I have included in the group Aphetohyoidea, is much less direct and conclusive. In very few of them are the branchial and hyoid arches ossified; indeed even the jaw cartilages are seldom visible, even though their position is made clear by the nature of the membrane bones which overlie them. In the arthrodire *Deinichthys* there are two ossifications in the palatoquadrate cartilage. One lies anteriorly held between the postsupragnathal cutting blade and the lower surface of the suborbital below the eye; the posterior bone is essentially a quadrate. In the lower jaw there are also two cartilage-bone bones, one corresponding to an articular (the so-called posterior infragnathal), the other quite anteriorly forming a symphysis with its fellow. These four bones are clearly homologous with the corresponding elements in *Acanthodes.* No traces of actual hyoid or branchial arches are preserved, nor is the neural cranium ossified in the ear region where a hyoid arch might be expected to articulate. But the whole structure of the dermal bones of the skull, and of the shoulder girdle, makes it evident that the gill-bearing region was short from back to front, and that to secure an adequate respiratory surface the gill slits must have been very long from top to bottom. In the somewhat aberrant arthrodire *Rhamphodopsis trispinatus* the palatoquadrate is ossified apparently as one element; lying behind it, parallel to the posterior margin, are the bones in the upper part of the hyoid arch. There is quite definitely a small pharyngohyal, and a short, very slender epihyal below it, which ends some distance dorsally to the articulation of the upper and lower jaws. Finally the arch is shown to possess an ossified ceratohyal, though nothing is to be seen of any more ventral element. It is evident that this arch does not contribute to the support of the jaws as a hyomandibular does. It is a continuous and independent chain of cartilages, in part ossified, and the material shows no reason why a large gill slit should not have been present in front of it.

But quite recently Professor Stensiö, in a lecture given in Paris, exhibited figures of the still more aberrant animal of arthrodire affinities, *Jagorina*, in which he claims to see an actual hyomandibular-like connection between the single upper element of the hyoid arch and the palatoquadrate. I have not myself seen the material, but it is not improbable that this hyostylic condition really existed. *Jagorina* is closely related to *Gemündina,* and *Gemündina* is a dorsoventrally compressed skatelike fish which possesses, as I have pointed out (Watson, 1937), a protrusible mouth. Such an arrangement is paralleled by numerous fishes including, in a somewhat modified form, certain skates, and I think that the hyostylic condition, if indeed it really exists, has arisen as an adaptive device in association with this protrusibility. In any case it is quite evident that *Gemündina* is an operculate fish possessing a large gill chamber. It is evident that conditions existing in *Gemündina* have no direct bearing on those to be seen in *Acanthodes* because no one can dispute that the hyostyly of elasmobranch and bony fishes must be secondary to the establishment of jaws.

The segmentation of the head in craniate vertebrates has been a subject of study since the classic Croonian Lecture by Huxley in 1858 on "The Theory of the Vertebrate Skull," which, by freeing the subject from the theoretic views of Oken and Owen, laid the whole matter open to investigation. The whole modern theory derives mostly from the embryological work of F. M. Balfour. Essentially the argument is as follows. In all craniate vertebrate animals the first sign of body segmentation is the cutting up of the originally continuous strip of paraxial mesoderm, which lies adjacent to the notochord, into a series of blocks of which each appears later in time than that lying in front of it. From the somites so formed arise the segmental body musculature and the segmental axial skeleton. As it is evident from inspection of the protochordates belonging to the Tunicata and the Enteropneusta that such a segmentation is not a primary quality of the chordate stock, while the other group of protochordates, *Amphioxus,* possesses an even more complete segmentation of the mesoderm, it is reasonable to assume that this segmentation was introduced to serve some function, as evidently the analogous segmentation of annelids was. The purpose can only have been to enable the primitive segmented chordates to swim by lateral flexion of the body. Such movement involves the contraction of segmented muscles at definite time intervals from front to back, the muscles of opposite sides necessarily being in opposite phases of contraction. For such conditions to exist a nervous system must be provided which will stimulate each muscle block to contract at the proper time interval after that which lies in front of it. There must in fact be segmentally arranged nerves passing to each myotome, and these nerves must contain both motor and sensory components. Furthermore the central nervous system must include longitudinally running fibers which will link each segment with its neighbors before and behind. Thus the segmental arrangement of spinal nerves is secondary to that of the muscles. It seems clear also that a similar segmental arrangement will be imposed on the skeletal and vascular systems; the fact that the dermis is derived from the segmentally arranged somites insures that the general cutaneous sensory nerves will have a segmental arrangement though those of the sympathetic system which innervate the blood vessels need only have a segmental arrangement insofar as the structures they supply are so placed. But convenience of arrangement in securing an unimpeded passage for nerves and blood vessels between the elements of the segmental muscular and skeletal elements is likely to induce a segmental arrangement even of other nerve components, and in fact we find in so primitive a vertebrate as *Petromyzon* that the dorsal and ventral roots of the spinal nerves do not fuse to form a common trunk. This segmentation of the body, which is shown in its simplest form as the production of somites from the paraxial mesoderm, would be expected to continue on into the head, and it is the great merit of van Wijhe to have shown that it does so. A somite in vertebrates normally possesses a small cavity, the myocoele, around which the cells are arranged in a somewhat radiating manner. Van Wijhe was able to show that far forward in the head of an elasmobranch embryo there exist three pairs of spaces, differing widely in size and each surrounded by a thin sheet of cells, and that from the walls of these three cavities the six eye muscles were formed by a process closely comparable to that by which ordinary body muscles are formed from normal somites; he pointed out that the three eye muscle nerves, the third, fourth, and sixth, found in every craniate which possesses eyes, supply the first, second, and third of these sets of eye muscles. He therefore concluded that these so-called head cavities in elasmobranchs are indeed somites, and that hence the segmentation of the head of craniates does in fact extend forward through the eye region in these animals. Much later it was shown by a long series of authors that the eye muscles of nearly all other craniates had the same mode of origin.

Meanwhile it was shown by Koltzoff in 1899 that in lampreys there is a complete, continuous series of obvious somites extending back into the trunk from the first one, from which are formed the eye muscles innervated by the third cranial nerve, so that in this creature the head mesoderm, except for that which lies around the nasal capsule at the extreme anterior end of the head, is completely segmental.

This arrangement is reflected in the condition of the segmental nerves. All muscles formed from somites are innervated by nerves which come out from the ventral side of the central nervous system, and of these the eye muscle nerves are the first three. The dorsal roots were not primitively connected with the ventral roots, and in the head region of all craniates this primitive nonassociation has been preserved. Their mutual spacial relations show that the sixth nerve has as its dorsal-root equivalent the seventh facial cranial nerve, and that the so-called fifth cranial nerve, the dorsal root of which lies in front of the seventh root, is really double, the ganglia of a trigeminus and a more anterior profundus nerve having coalesced to form the Gasserian ganglion of man and of most craniates. Thus the trigeminus proper is related to the fourth cranial nerve, the profundus to the third. These conditions imply that somites primitively existed, presumably as a continuous series, from the region of the profundus backward, that is from the anterior end of the notochord, the point at which the anterior end of the head process of the developing embryo becomes indistinguishable from the endoderm which there roofs the anterior end of the fore-gut.

It must now be considered how far the segmentation secondarily imposed on the body extends; whether in fact the piece of tissue which forms the actual anterior end of the animal is to be regarded as a true body segment or whether it is a piece left over, as it were, in the process of division. The extreme anterior end of the body of any bilaterally symmetrical animal is that which first comes into contact with outside bodies or influences as the animal moves forward. It is therefore likely to be especially well provided with sense organs of various kinds, and it may very well have had special peculiarities, of significance in the intake of food or in some other activity of an animal which was certainly mobile, and no fixed form. Any musculature contained in such a structure would be related to some specific requirement of this specialized region of the animal, and it is in every way improbable that it could take part in that cutting up of the muscle system into a repetition of short blocks which is the essence of chordate segmentation. In other words an anterior piece of the animal, provided with sensory but not necessarily with motor nerves, remained outside the segmented region of the body. Two structures have been found in craniates which may be related to it. The nervus terminalis is unique in that the cell bodies of the neurons which provide its fibers lie scattered irregularly along the length of the nerve. It is clearly sensory, it passes toward the extreme anterior end of the body, its function is unknown but its fibers do pass into the extreme morphological anterior end of the central nervous system. It is also possible that the anterior head cavity, described by Miss Platt in *Squalus,* which gives rise to no structure in the adult, may possibly represent a coelomic space developed in this anterior region.

All protochordates are microphagous feeders; they obtain their food by sieving out from the sea water in which they are found minute living beings, algae or other small plants, or the detrital remains of such organisms. Thus their structure must be modified to serve this end. The most simple type for discussion is probably *Amphioxus.* Microphagous feeding, if it is to be adequate, depends on the sieving of a body of water relatively large in proportion to that of the animal which it is to supply. The evidence afforded by sponges and lamellibranchs suggests that the most effective way of treating so relatively large a body of water is to set up a current coming in through an inhalant aperture, passing through a sieve, and expelled through an exhalant opening. In the protochordates, *Amphioxus,* and the tunicates, this is done by ciliary action; the inhalant opening is the mouth and presumably always has been the mouth; the exhalant opening is provided by a series of perforations connecting the cavity of the gut with the exterior and thus penetrating not only the wall of the gut but the body wall outside it. This implies the formation of a series of tubes in linear succession, whose epithelial linings will be formed in part from the endoderm and in part from the ectoderm. The nature of the current is evidently of importance to the animal, and chemical and tactile sense organs may be expected to make their appearance somewhere in the system. These openings are gill slits, and the fact that they occur in vast numbers in the unsegmented tunicates probably merely depends on the fact that it is possible to provide a sieve only across a relatively small aperture, and only by multiplication of such apertures can the necessary large amount of water be allowed to escape. Thus at its origin the apparent segmentation of the gill slits had a functional significance of its own, unrelated to the much later segmentation of the body muscle system. It seems likely that the mouth, the point of entrance of water, would lie immediately below the presegmental anterior end of the body, for there it would receive water whose nature could have been determined by the sense organs, which we have seen reason to believe would have existed on the prostomium. When by segmentation of the musculature the locomotion of the animal became more powerful, perhaps concurrently with an invasion of estuaries and fresh waters, the original feeding mechanism with its series of gill slits, the pores we have already discussed, would survive with little modification, as indeed it does in the ammocoete larva of *Petromyzon*.

The muscle segmentation, if we may judge from the course it pursues during the development of a craniate, was originally restricted to a strip along the dorsal side of the animal. Such an arrangement is unsound mechanically and it is evident that a rearrangement whereby the myotomes extend from the mid-dorsal to the mid-ventral line is necessary if the animal is to be able to pursue a straight course through the water. Such a ventral extension necessarily brings a segmented musculature into the region perforated by the gill slits, and it is obvious that the only possible place for the gill slits is between the myotomes, in effect through splits in the myocommata. Thus the segmentation of the body and the head imposes itself on the much older multiplication of gill slits, and as a result the senses of the nature of taste, which are associated with the gill slits, must be served by splanchnic nerve fibers which run in the ordinary segmental dorsal nerves.

Although some ciliary-feeding animals attain a large size they do so only in the case of sedentary animals whose food requirements are relatively low. When the vertebrates became actively moving, at the time of the appearance of the segmental musculature, they must have achieved a much increased metabolic rate. Their need for food and for oxygen was greatly increased. Oxygen is easily dealt with, assuming that by the introduction of hemoglobin the oxygen- and carbon-dioxide-carrying power of the blood be sufficiently increased. The only other requirement is the provision of a sufficiently large respiratory surface exposed to water to allow of an adequate inflow of oxygen by diffusion. This in turn demands the reduction to a minimum of the distance between the external surface and the lumen of the superficial capillaries. The answer is the establishment of a gill, essentially a much subdivided mass of folded skin whose surface area is very large in comparison with the base to which the whole structure is attached. The thin epithelium and its immediate proximity to capillaries make such a structure highly liable to damage. It must therefore be placed in a protected place; the fact that the gill clefts passing through the musculature have become of considerable length enables gills to be attached to their walls and thus possess an adequate surface area, in a position where they are not only protected from the liability of mechanical damage but also so placed that they are constantly bathed in a stream of water taken in directly from the mass in which the animal is swimming.

This new respiratory use of the gill pouches imposes certain qualities on a vascular system. The gill filaments must be supplied with deoxygenated blood, and the oxygenated blood coming away from them must pass to the active organs of the body. Hence arises a segmentation of the anterior part of the blood system for reasons other than that which brought into being the normal, somatic, segmental blood supply. The position of the heart is fixed by that of the last gill slit, and it is obviously undesirable that it should lie very posteriorly in the body, so that one would anticipate a reduction in the number of gill clefts; it is evident from the whole of the groups of lower craniates that such a reduction has occurred, the hinder end of the series moving progressively farther and farther forward. With this reduction in numbers the amount of water which could be pumped through the pharynx by ciliary action would clearly become less and less because the area of effective ciliated epithelium would almost necessarily have been reduced. Some other mode of securing the persistence of a current therefore becomes necessary, and it is evident that alternate constriction and expansion of the pharynx brought about by muscular movements is one such method, the alternative being the introduction into each gill cleft of a muscularized expansion which, if suitable valves were provided, would serve as a pump.

The earliest animals, whose structure is known with any completeness, representing this grade of structure are the cephalaspids, where the classic and altogether admirable work of Professor Stensiö has given us a knowledge of the anatomy of extinct forms incomparably more complete than we could have expected ever to attain. According to his description, the more important points of which are entirely confirmed by examination of the British collections, *Cephalaspis* and its allies are fresh-water, bottom-living animals in which the head, branchial region, and a variable length of the anterior part of the trunk are surmounted and surrounded by a co-ossified series of dermal bones. From this head shield, which is produced laterally, often into long spines, the rest of the body arises abruptly. It is triangular in transverse section, the flat ventral surface being connected with the slightly smaller lateral surfaces by a row of angulated scales; a similar series forms the mid-dorsal ridge, which is drawn up locally into one or two dorsal fins. Posteriorly there is a heterocercal caudal fin. From the hinder surface of the head shield, laterally to the insertion of the body, arises a pair of highly peculiar pectoral fins, scale covered, with an internal musculature, and presumably, although there is no direct evidence on the matter, containing a skeleton. Thus the head shield ends posteriorly in a bony wall which is in effect a shoulder girdle, and shows that the whole branchial region lies farther forward below the shield. In many cases the ventral surface of the head shield, onto which turns a dermal skeleton continuous with that of the upper surface, is perforated by a large opening with a continuous margin. This border is notched anteriorly, where there is a relatively narrow bay, behind which small notches symmetrically placed about the middle line extend back to a number of ten pairs. The infilling of this opening has never been completely seen in a typical *Cephalaspis*. It is best shown in *Micraspis gracilis* Kiaer and in *Hemicyclaspis murchisoni,* but even here the structure is not fully exposed. Unfortunately the material does not make clear the relation of these plates to the notches in the border of the ventral opening. Fortunately the late Professor W. Patten in 1901 had already shown that in the Lower Ludlow *Tremataspis* the corresponding notches are completed by the series of relatively large dermal bones which in this animal fill the aperture so that they form two rows of openings, nine or ten in number on each side, which are quite clearly essentially circular openings passing inward and can only be gill slits similar to those of the lamprey. Rohon had earlier recognized their position and the conclusion seems to me inescapable. Anteriorly *Tremataspis* possesses a mouth, the median point of the ventral border of the head shield projecting backward as a point which, in a specimen Professor Patten showed me in 1932, opposes the anterior and inner ends of a pair of relatively large plates in the anterior part of the ventral squamation.

It seems to be impossible to determine from any material hitherto described the actual position of the corner of the mouth, although its general character as a narrow, transversely placed opening whose lower lip was movable is obvious. Thus *Cephalaspis* has a jawless mouth and a paired series of about ten gill slits. Professor Stensiö in 1927 showed that below the dermal bone of the head shield in several cephalaspids there lies a very extensive endocranium whose nature must now be discussed. This structure contains far more than a neural cranium and is of entirely peculiar nature. It is continuous throughout, and it is impossible to show that it ever consisted of separate elements. In general the bone in this endoskeleton usually consists only of an excessively thin lamina which coats surfaces. It forms, for example, continuous bony canals surrounding the cranial nerves and their branches almost from origin to final distribution. It coats canals through which blood vessels pass, and it is limited ventrally by a great, continuous, smooth sheet of bone which extends over the whole of the very widened pharynx. It is so complete that it may fill the cavity of the cornua of the head shield, but in general there is nothing whatever in the way of bone connecting these superficial laminae. As Stensiö has pointed out, the only possible interpretation of this most unusual condition is that the bone was laid down as perichondral bone on the surface of an extensive mass of a tissue sharply marked off from ordinary connective tissue, which can only have been some form of cartilage, perhaps the mucocartilage of the ammocoete larva of *Petromyzon*. The lower surface of this endocranium, as is demonstrated especially well in a number of wax-plate models made from ground sections of specimens of cephalaspids, has a shape which evidently was modeled on the branchial and pharyngeal region of the gut and on the efferent vessels and dorsal aorta which were associated with them. Each one of the notches in the ventral border of the head shield is shown by these models to represent the outer end of a groove which passes inward toward the mid-line. In the posterior region of the head shield immediately in front of the shoulder girdle these grooves are, in *C. hoeli*, separated from one another by deep ridges, and it is immediately evident that each of the whole series of such grooves, and the ridges which separate them, is the homologue of every other. The obvious suggestion is that they are associated with the gills. The nature of these gills, however, is very far from evident. The entire dorsal part of the structure is rigid and quite immobile. What the ventral part was like we have no evidence, all we know is that it rested ultimately on a floor whose lower surface is covered with a series of relatively small dermal ossicles, not fused with one another and therefore presumably flexible. From a point far forward in the head shield the mid-region of this ventral bony sheet above the pharynx is drawn downward into a ridge, which is grooved. This groove evidently clasped a dorsal aorta, which in some forms actually passes down into the ridge, so that it lies in a canal perforated by foramina leading into lateral transverse grooves, which can only have been formed around efferent branchial arteries. In one form or another these can be seen to form a quite continuous series from the hindmost to the most anterior of those paired depressions, each of which ends laterally in a gill slit. There are, as one would expect, some cases of fusion be-

tween the admesial ends of these arteries, but their serial homology
is perfectly evident. Into the lateral ends of each one of these grooves
housing the gills there opens a foramen which often leads into a
short groove which then soon forks into two. These foramina lie at
the ends of bony tubes leading directly to the brain cavity which,
beyond dispute, housed branchiomeric cranial nerves. There are in
general six divisions of the vagus. The ninth cranial nerve is largely
independent; it has its own root, but its ganglion is closely associated
with that of the vagus. The seventh nerve arises in its normal position
and corresponds, even in its method of branching, with the ninth.
The trigeminus has precisely the same character and its root is far
removed from that of the facialis; and the ophthalmicus profundus,
entirely separated to its root from the trigeminus, has been shown
by Stensiö in a very elaborate discussion to be identical in nature
with the trigeminus. Indeed the relation of its distal extremity to the
groove on the lower surface of the endocranium on which it opens
is, in principle and very largely in detail, indistinguishable from that
of the trigeminus, the facial, or the glossopharyngeal. It seems therefore evident that the whole series of grooves was occupied by a corresponding series of gills whose nature must now be considered.

In *Kiaeraspis* the bony shoulder girdle which lies behind the
branchial region is perforated by two large holes, the more ventral
of which is evidently for the passage of the ventral aorta, while the
more dorsally placed (which lies below the level of the dorsal aorta)
must have transmitted the esophagus and little else. It is of good size
but is only about one-quarter of the width of the branchial region
at its widest. If now we consider the width of the most anterior part
of the pharynx, where it joins the oral cavity, we are driven into
consideration of the width of the mouth. This cannot be directly determined, even in *Tremataspis* where the whole of the ventral surface is most clearly preserved, but Stensiö's models of this region
in *Kiaeraspis* and in *C. sp.* (Stensiö 1927, text Fig. 13) show that
there is in the mid-line a triangular area whose base lies anteriorly
on the lower margin of the head shield, which is quite clearly marked
off from the first branchial depression by ridges, each grooved for
a blood vessel formed essentially by a splitting of the anterior end
of the dorsal aorta. This area is not shown to receive any direct
supply from the head nervous system; it appears to have a rather
independent venous drainage shown in many specimens of many
genera, and it is clearly the actual oral cavity, the velum (if it still
survived) being borne by the ridges which bound it. The maximum

width which this mouth can have possessed is of about the same order as the width of the foramen through which the esophagus passed. We have therefore to consider the width of the continuous pharyngeal cavity in the branchial region. There is in fact nothing in the described material which gives any indication of the width, and the matter can be discussed only on the basis of probabilities or possibilities. We know from the evidence of Professor Patten's specimen of *Tremataspis* that the ventral border of the mouth in that animal was supported by certain dermal plates, of which those that lay in contact with one another in the mid-line are long from back to front. Furthermore in this specimen there seems definite evidence suggesting that they were separately movable with respect to one another in that they seem to show signs of wear on their admesial and anterior surfaces. In *Micraspis,* as shown by Kiaer, a series of elongated but nonetheless small dermal plates border the ventral side of the mouth, but as to their nature nothing can be said. These elements recall those which White has described in a similar position in *Pteraspis,* and it is interesting to speculate as to their function.

The shape of *Cephalaspis* shows beyond question that the creature rested on and probably generally moved along the bottom. Its mouth is entirely on the ventral surface and it must therefore have obtained its food from the bottom. There are thus two possibilities. We may exclude any conception of a diet of plankton; but the surface of the floor of any considerable body of water is usually full of life—bacteria, protozoa, and various metazoan animals—and it is conceivable that the mouth of *Cephalaspis* is of such a character as to fit it to scrape off the superficial layer. The other possibility, which seems to me on the whole less likely, is that it fed by swallowing relatively large quantities of sand or mud and digesting out the food particles that it contained. In either case the water entering the mouth, which actually forms the respiratory stream on which the animal depended for oxygen, is likely to have been exceedingly muddy, a condition that would require the development of adequate means of keeping the surface of the gill filaments clean.

It is difficult to conceive a sorting mechanism existing in the pharynx which would separate food from inorganic particles and send one down to the esophagus and expel the other, so that in any case we might expect to find considerable quantities of mud passing down the gut. A ciliary mechanism carried on the gill filaments might be adequate to keep them clean, and may have existed, but it is difficult to imagine that it could ever have been adequate to draw in through the mouth a sufficiently large volume of water, so that we must postulate the existence of a muscle system capable of rhythmically increasing and decreasing the volume of the pharynx and of the gill clefts. The floor of the pharynx seems certainly to have been movable; what was the nature of any visceral skeleton which lay in it we do not know. It cannot conceivably have been similar in structure to the branchial arch skeleton of a gnathostome, and the most obvious form of musculature would be that of a sheet crossing the opening of the ventral side of the head shield. Such a muscle, however, being flat when contracted, is very disadvantageously placed, and it is far more likely that the gill clefts of *Cephalaspis* expanded into pouches similar to those of the living marsipobranchs and that each of these had its own independent musculature.

We may therefore conceive that each of the paired concavities in the continuous sheet of ectochondral bone which lies over the pharynx contained a gill pouch, whose wall was muscular, or was at any rate actuated by muscles, the pouch being connected with the pharynx and with the external surface of the animal by cylindrical tubes. The volume changes in the head which would accompany the expansions and contractions of these pouches would be made possible by the flexibility of the floor of the head. The alternative explanation, that volume changes in the pharynx were brought about by muscular movements of a pharyngeal skeleton, is very difficult to believe. The floor of the pharynx might in principle be raised by muscles attached to it stretching vertically to arise from the undersurface of the head shield, but it is not easy to conceive how adequate opponents of these muscles could be provided.

Additional evidence that *Cephalaspis* possessed pouch gills is given by the fact that the external gill slit is circular in plan in *Tremataspis,* may be (and probably is) circular in *Cephalaspis* itself and its immediate allies, and is certainly circular in the whole group of Anaspida, which is evidently closely related to cephalaspids. Additional evidence may perhaps be drawn from the not closely related Heterostraci, where in *Pteraspis* and all its allies not only the upper but also the lower surface of the pharyngeal region is held rigid by powerful plates of dermal bone. In these animals the gills discharge, not through independent gill slits but through a common exhalant aperture lying far back, and it seems quite obvious that no branchial skeleton and musculature analogous to that of true fishes can possibly have existed, so that the only possibility is the existence in these

forms of pouch gills of cyclostome pattern. The conditions existing
in such a form as *Cyathaspis*, where the undersurface of the dermal
head shield (which in these forms does not include an endoskeletal
portion) recalls quite closely the appearance of the ridges and grooves
on the undersurface of the endocranium of a cephalaspid, lend some
further support to the view which I have here adopted.

Thus there are some reasons for believing that typical marsipo-
branch pouched gills existed in cephalaspids, and the details made
out by Stensiö of the relation of the cranial nerves to the cavities in
which such pouches must have lain are perfectly consistent with
this view. But any impartial consideration of Stensiö's text Figures
4, 13, 28, and 33 of his 1927 paper will show that the most anterior
of the depressions, which is innervated by the profundus nerve, differs
in no way from that innervated by the trigeminus which lies behind
it, or indeed from the rest of the series. Thus we must face the position
that the cephalaspids possessed two complete, apparently unmodified
gill pouches in front of the hyoid cleft, innervated by the seventh
nerve: that is, that the trigeminus and the profundus originally supplied complete normal branchial clefts, the mouth lying still farther
forward; and that *Cephalaspis* shows us in an adult craniate conditions which have or might have been predicted from the existence
of the profundus and third cranial nerves as homologues of the
seventh and sixth cranial nerves, which are related to a gill slit, the
spiracle. It seems quite obvious that the anaspids had already lost
their anterior gill slits in the adult, so that even in those very archaic
forms, which may well be of Ludlow or even Wenlock age, modification of the anterior end of the body for special purposes had already
gone on.

II. UNIFORM CHANGE

In this chapter I propose to show by discussion of an actual case how morphological methods are used in sorting out series of animals which are then built into phylogenetic trees.

This process involves the determination of the order of appearance in time of the animals which form part of the various phyletic series, and it is necessary to understand, at least in outline, the basis on which geologists have built up a time scale. Fundamentally the argument runs on the following lines. There exist many rocks which contain fossil shells, bones, or plants so well preserved that no unprejudiced person can doubt that they were once parts of living organisms. Indeed, by hunting in the sediments laid down by recent floods, rivers, and raised beaches, in which the remains of domestic animals are to be found in association with human artifacts, and so gradually passing back to prehuman times, this fundamental conclusion can be placed beyond dispute.

Many of the rocks in which fossils are to be found are still sands or clays identical with those of the sea floor or deserts. From such unconsolidated sediments it is possible to trace by continuous and unbroken series a transition to hard sandstones, shales, and limestones, to quartzites, slates, and marbles, and finally to the foliated schists and gneisses.

By direct comparison with existing conditions it is possible to determine the conditions under which any particular bed of sandstone or shale might have been laid down, and indeed a sufficiently extended series of observations may enable one to determine with a very high degree of probability the actual circumstances under which it did in fact come into existence. But for our immediate purpose all that is necessary is to realize that each layer or bed of sandstone or shale was at first a layer of sand or mud spread out on the surface of a pre-existing body of rock, usually no more than a similar layer of unconsolidated detritus, by the action of a natural agent, the sea, a river, or the wind. The layer of sand so laid down must be younger than that on which it rests. Such superposition is thus absolute evidence of succession in time. If all these sediments are compacted and preserved, they will for all time preserve the record of the relative time of formation, except in those very rare cases

31

where a series of rocks by inversion or by thrusts comes to rest on more recent deposits. But these cases are relatively rare, and as they necessarily occur in regions of intense compression are so easily recognized that they do not really render the establishment of a general stratigraphical succession difficult.

It is thus possible, through application of the principle of superposition and by a process of mapping, to determine the relative ages of all sedimentary rocks which come into contact with one another. The process is so simple in its essence, and even sometimes in its application, that in 1790 William Smith, an inconspicuous land surveyor, was able to show that in England the order of succession of different groups of rocks is never inverted and that these groups might be identified at very distant points by the peculiar organized fossils which they contained. In 1815 William Smith, who by then had traveled the whole of England on horseback and on foot, had published a geological map which was entirely correct in its broad outline. England is an admirable country for such a study because it is possible to walk from Dorset to Yorkshire, some 250 miles, along a single geological horizon and so convince oneself of its continuity, and yet in the 250 miles from London to Wales to cross rocks which cover very nearly the whole of the geological record.

Thus in England the order of succession is a matter of direct observation, and the second of William Smith's generalizations could be established by sufficient collecting and accurate determination of fossils. That it was completely justified has been shown by the work of geologists in all parts of the world during the past century. That the age of a bed can be established by the fossils it contains is the fundamental fact of paleontology, and is indeed essential to my present purpose because it provides not the only, but by far the most generally applicable, method of comparing the ages of sediments in regions as far removed from one another as North America and Europe.

To William Smith and to his successors until the time of Darwin the doctrine was purely empirical; the value of the evidence grew with the collection and determination of each fossil, until now that fossil faunas have been collected from every part of the world, and in no case has the order of succession been inverted, it has become as firmly based as any such generalization can ever be. And the existence of this world-wide order of succession of faunas is the strongest of all evidence for the occurrence of organic evolution.

Thus we can accept, as an empirical fact and as one which is con-

sonant with our theoretical views, the possibility of determining
the contemporaneity of definite groups of stratified rocks in widely
separated parts of the earth's surface.

During Devonian, Carboniferous, Permian, and Triassic times
there were continental regions in which were laid down great series
of sedimentary deposits, which by subsequent denudation have become visible and exposed to the attentions of geologists.

The deposits of Devonian, Permian, and Triassic times are, in
many parts of the world, of a nature which suggests that they were
the product of semi-arid countries, deposited by seasonal rivers along
their banks, and in impersistent lakes liable after a succession of dry
seasons to become very small or to disappear. Such rocks are to be
found all over the world—in Greenland, North America, Brazil,
South Africa, Europe, China, India, and Australia.

In Carboniferous times few or no rocks with such an origin are
known, but during this period, in the Lower Carboniferous of Scotland and in the Upper Carboniferous of England, Czechoslovakia,
and the United States, we do find continental deposits of the very
peculiar character of Coal Measures. These were laid down on low-
lying coastal areas covered intermittently in any given area by vast
forests lasting only for what was in a geological sense a very short
while. They were intersected by slow-moving, meandering streams,
occasionally widening out to form shallow lakes in the woods a mile
or more across. The muds laid down in these ponds usually contain abundant fragments of fish and other aquatic animals.

Both series of rocks, those of semi-arid deposition as well as the
Coal Measures, yield the remains of animals which can be recognized by the presence of a characteristic ornament on their skulls,
lower jaws, and some other bones (Fig. 10). These bones have a
smooth inner surface, while the exterior is excavated into a series
of pits and grooves separated from one another by ridges whose
surface is smoothly rounded. Each bone has an initial area in which
the pits are essentially circular, while in most of them the periphery
is covered by more elongated grooves and ridges so arranged that
they radiate from the center of this initial area, the length of the individual groove being related to the extension of the bone along
that direction. The skulls which bear this ornament vary greatly in
shape (Fig. 11). They may be long and sharp snouted or extremely
broad; they may be high or very flat; they may have a smooth outline
or be produced into horns; and those of adults may vary in length
from about one centimeter to giants of one meter.

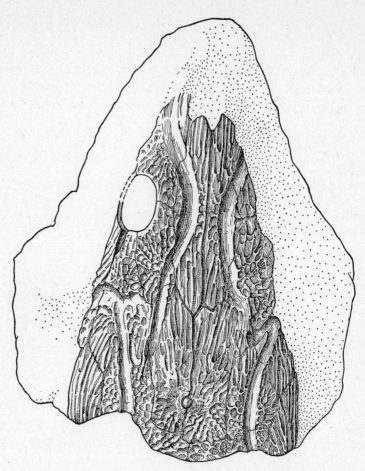

FIGURE 10. The upper surface of the skull of the labyrinthodont *Trematosaurus*. This drawing shows the ornament, a sculpturing of the outer surface of the dermal bones into a series of pits, nearly isodiametric over the center of the individual bones but elongated into grooves and ridges in those regions where growth has been most rapid, which is characteristic of the Paleozoic amphibia although it does not occur in all the species within the group. The mid-line of the skull is vertical, and the anterior end uppermost. The left orbit is shown to the left, and the upper border of the right orbit is symmetrical with it. The continuous wide grooves are for lateral-line canals. The drawing was made from a plasticene squeeze of a mold in a block of sandstone from the Middle Bunter of Bernburg, Thuringia, now in the author's collection.

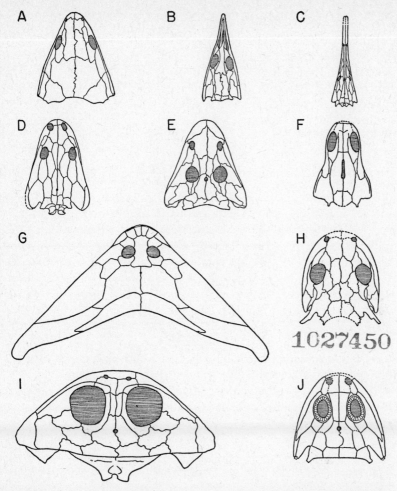

FIGURE 11. A series of drawings of the upper surfaces of the skulls of Paleozoic amphibia, to illustrate the extreme variety of form to be found among them while the fundamental pattern made by the membrane bones remains constant in the sense that all may be derived simply from a single fundamental type. All the drawings are reduced to the same length from occiput to snout.

A. *Microbrachis pelikani*, Upper Carboniferous, Nyrany, Czechoslovakia;
B. *Urocordylus* sp. Upper Carboniferous, Linton, Ohio;
C. *Lonchorhynchus*, Middle Trias, Spitsbergen;
D. *Buettneria*, Upper Trias, Texas;
E. *Nyrania*, Upper Carboniferous, Nyrany;
F. *Adelogyrinus*, Lower Carboniferous, Scotland;
G. *Diplocaulus*, Lower Permian, Texas;
H. *Stegops*, Upper Carboniferous, Linton, Ohio;
I. *Plagiosaurus*, Upper Triassic, Württemberg;
J. *Pelosaurus*, Lower Permian, Czechoslovakia.
A and F are Adelospondyli; B and G are Lepospondyli; C, D, E, and I are labyrinthodonts; H and J, Phyllospondyli. A, B, E, and H after Steen; C after Säve-Söderbergh; F from Watson; G from Williston; I from Fraas; J from Bulman and Whittard.

Further inspection of these skulls will show that they are composed of a skin, built up by a large number of bones rigidly attached to one another by sutures, which covers the whole of the top and sides of the head, leaving only openings for the nostrils, eyes, and pineal organ, and is notched behind for the tympanic membranes of the ears. The dermal skull is attached to the brain case in a way that we shall consider later.

By good fortune it happens that in some places, for example in the Lebach Ironstones of the Saar, skulls of this kind are common fossils and with few exceptions all belong to one species. *Archegosaurus* from this place, the first labyrinthodont amphibian ever found (1847), is known by hundreds of skulls of all sizes from about two to thirty centimeters in length. The detailed examination of this material by H. von Meyer in 1857 showed that the pattern formed by the many individual bones was constant, in the sense that the same number of bones was to be found in every skull and each individual bone had constant relationships with its neighbors. As the skull shape varies considerably, the snout lengthening as the animal grows older, the shapes of the individual bones necessarily vary systematically with age. Dr. Margaret Steen, working in my laboratory, has shown that the same constancy and similar age changes occur in species of very different head shapes.

If we endeavor to compare together skulls of very different shape we soon find that it is difficult to determine directly the homologies of individual bones, but that even in these cases it is evident that rows or groups of bones can be compared directly. By selecting suitable series of animals the gaps can be made so small that it becomes obvious the fundamental skull pattern is constant; even the most extreme forms are reducible to it by at most the reduction and ultimate loss of one or two bones, the bones so lost being always the same. Thus by applying the criteria of ornament and skull pattern it is possible to sort out from a much greater bulk of material skulls representing more than a hundred genera. For a considerable proportion of these we know something of their skeleton, and in some cases we know the whole animal very completely.

It is evident that before we can go further with our analysis we must break up the bulk of material into smaller groups. The most convenient way of doing so is to consider the structure of the vertebral column (Fig. 12). The largest forms, the true labyrinthodonts, have vertebrae consisting of neural arches which span the spinal cord and in the back are carried out into transverse processes to the

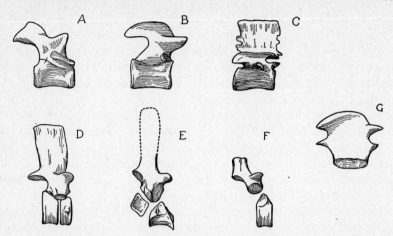

FIGURE 12. A series of vertebrae showing the various types occurring in Paleozoic amphibia. All seen from the right side (the head of the animal being to the right).
A. *Hylonomous* sp., Upper Carboniferous, South Joggins, Nova Scotia;
B. *Lysorophus* sp., Lower Permian, Texas;
C. *Crossotelos* sp., Lower Permian, Oklahoma;
D. An embolomerous labyrinthodont, Upper Carboniferous, Airdrie, Scotland;
E. *Eryops*, Lower Permian, Texas;
F. *Cyclotosaurus*, Upper Trias, St. Peters, New South Wales;
G. *Branchiosaurus*, Lower Permian, Germany.
A and B are Adelospondyli; C a lepospondyl (nectridian); D, E, and F the labyrinthodont line; G is Phyllospondyli. Fig. F from a specimen in the British Museum; the remainder from the author's collection.

ends of which the tuberculum, a special projection of the rib, is attached. Lying below the neural arch is a variable series of bones: in early forms two biconcave discs with a central perforation. In later animals of this group we find one horseshoe-shaped central bone and a pair of more dorsally placed quadrangular bones. In the last we have only an enlarged horseshoe-shaped intercentrum.

These three types are so closely linked up to one another by intermediate stages that it is evident they represent variants of a common fundamental pattern. The group characterized by the possession of these vertebrae is called the Labyrinthodonta, and the three types of vertebrae are called the embolomerous, with disc-like pleuro- and intercentra; the rachitomous, with a wedge-shaped intercentra and paired pleurocentra; and the stereospondylous, with only a single horseshoe-shaped intercentrum. These names of vertebrae are used to characterize evolutionary stages of the labyrinthodonts.

The next type of vertebra consists only of a neural arch, built up by a pair of independent bones, which on transverse processes supports short ribs and rests upon unossified structures below the spinal cord. This phyllospondylous vertebra is characteristic of another group of fossil amphibia, but it occurs transiently as a stage in development of rachitomous and stereospondylous vertebrae.

The third type of vertebral column found in connection with skulls of the kind we have sorted out is that called by von Zittel lepospondylous. Here the whole vertebra is a single ossification, the neural arch, even in young individuals, being continuous with the central part which surrounds the notochord. Neighboring vertebrae interlock by an unusually elaborate series of articular facets, and in some of the elongated eel-like forms which form one division of this group the neural spines are connected to one another by the intergrowth of processes like those of a suture between membrane bones.

The fourth distinct type of vertebra is that which I have called adelospondylous. In these the neural arch is always distinct and there is a single centrum, containing an hourglass-shaped cavity, which surrounds the notochord. This centrum is at least as long as it is wide and cannot be confused with any element found in other groups.

By applying the criterion afforded by the structure of the vertebrae in those cases where the preservation of the skull and body together makes it possible, the whole group of early amphibia can be split up into subgroups. Direct comparison of the patterns of the dermal bones of the head in members of each of these groups soon shows that each has its own characteristic version of the fundamental pattern. The Adelospondyli, for example, can be distinguished because in them alone the bone called the supratemporal is usually completely lost, while in all other groups it is retained as an important element. The differences in skull patterns are often small, but when experience makes one familiar with its nuances it becomes possible to attribute skulls to their proper order, even without knowledge of the postcranial skeleton.

Let us now consider the labyrinthodonts found in rocks of Permian and Triassic age. More than fifty genera are known by well-preserved skulls, many by considerable parts of their skeletons, and some completely. If we consider the range of structure which these fifty skulls present we find that it is possible to split up the whole series into groups on the basis of proportion and the position of the orbit. We can, for example, isolate a group with skulls of the shape I call capitosaurine after its first described member. In these forms the skull is parabolic with a wide, depressed snout, and the nostrils lie near to the margin at the anterior end of the skull. The orbits, within which lay the eyes, are placed behind the middle of the skull length and are relatively small and close together.

In contrast to this group we may select one in which the skull is triangular in plan, with only a very narrow transversely placed anterior truncation, and in which the snout extends in a long, slender rostrum. In many of such skulls the nostrils lie far back but are still lateral in position, and the orbits are placed quite posteriorly but still face upward rather than outward.

The opposite extreme is presented by the skulls called brachyopid. These are very short with a rounded contour, their width equaling or even exceeding their length. The nostrils are very small, placed close together at the extreme point of the head. The orbits are always large and may be enormous, and they face almost directly upward.

Still another group is distinguished by possessing flat heads with small, widely separated orbits facing directly upward and placed very far forward, not much behind the nostrils. These skulls are parabolic with a rounded contour and are longer than they are wide.

Another group contains a few animals with short, rounded skulls, flat on the upper surface, with their rather large orbits laterally directed and only a short vertical cheek behind them.

Finally there remains a number of skulls which do not fall into any of these groups.

If now we take the five groups into which we have divided the later labyrinthodonts and arrange the members of each in order of time, we find the very interesting fact that capitosaur skulls may be found in all horizons from the base of the Permian to the top of the Trias, while no other shape of skull has so extensive or continuous a range in time.

The elongated skulls are represented in the Lower and Middle Permian and in the Lower and Middle Trias. In contrast to them we find brachyopid skulls at all times from the Upper Permian to the Upper Trias. The skulls with very anteriorly placed orbits are found only in the Basal Permian and the Upper Trias, and those with laterally directed eyes in the Basal and Upper Permian only. Thus the only groups which can be expected to give the continuous history of a real evolutionary series are those of the capitosaurine and brachyopid types.

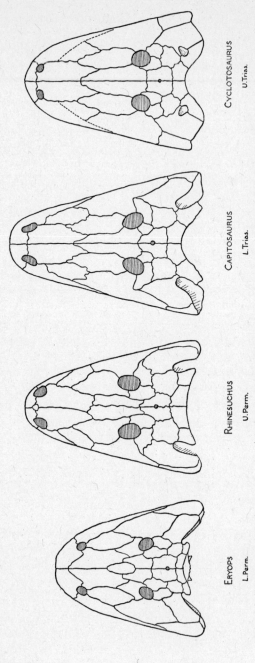

| ERYOPS | RHINESUCHUS | CAPITOSAURUS | CYCLOTOSAURUS |
| L. Perm. | U. Perm. | L. Trias. | U. Trias. |

FIGURE 13. A series of drawings of the dorsal surface of labyrinthodont skulls, selected merely on a basis of their generally similar shape and proportions. In all the orbits, which are transversely shaded, lie in the posterior half of the skull and are not very far apart; such skulls are called capitosaurine. The drawings are placed in the order of time of appearance of the animal, and the four forms have been picked from a much more extended series because suitable figures were available. The figures are reduced to the same width for ease of comparison. These figures compared with one another show only small changes in the skull pattern and confirm the view that they may be blood relations. *Eryops* after Broom; *Rhinesuchus* from a drawing by F. R. Parrington of a skull collected by him in Nyasaland; *Capitosaurus* after Schroeder; *Cyclotosaurus*, original reconstruction of *C. robustus* from Quenstedt's figures. The small median bone in the skull of *Eryops* is of inconstant occurrence and of no significance.

40

The earliest member of the capitosaurs is *Eryops* from the Basal
Permian of Texas, known by many skulls and several somewhat imperfect skeletons. The next form, *Onchiodon*, from the apparently
slightly younger Middle Rothliegende of Dresden, is less well known,
but the material fortunately includes skulls varying from juveniles
about four centimeters long to adults of forty centimeters, and some
skeletal fragments. Unsatisfactory fragments seem to represent the
group in the copper-bearing sandstones of the Urals. In the Karroo
systems of South and Central Africa the group is represented at
every named horizon from the base of the Upper Permian to the
top of the Lower Trias. These animals fall within the genera *Rhinesuchus*, *Uranocentrodon*, and *Capitosaurus*. Much of the splendid
material has been collected only recently and still awaits description.

In western Europe many forms of the Lower and Upper Triassic
genera *Capitosaurus* and *Cyclotosaurus* occur, and when these are
studied in connection with skulls from the Australian Trias it becomes clear that we possess a complete morphological transition
from one to the other genus. I intend to publish a description of this
material in the near future.

If we reduce a series of drawings of the capitosaurine skulls to
the same absolute width and arrange them in time order, we can
compare each with the next and assure ourselves that the pattern
made by the dermal bones, though not identical in the early and
late forms, is throughout on the same plan (Fig. 13). If we ignore
a small median bone in the roof of the skull of *Eryops*, which we
are justified in doing because according to Dr. Broom it is not of
constant occurrence within the genus, we find that the gap separating any stage from its neighbors is very small. But these differences
are cumulative, a fact which implies that there is a constant direction of evolutionary change throughout the group. It should be
pointed out that this condition affords strong evidence that these
animals are indeed blood relations. These changes in the dermal
roof of the skull are really small, so inessential that it is impossible
to determine the age of any specimen from them alone.

If, however, we turn the skulls over and consider their palates
we find obvious differences between them (Fig. 14). The palate of
Eryops is built up by membrane bones which arise in the skin roof-
ing the cavity of the mouth exactly as do the dermal bones which
roof the head. This structure is rigidly connected to the dermal
roof by the close interlocking sutures between its bones and those
which form the margin of the mouth and are themselves part of

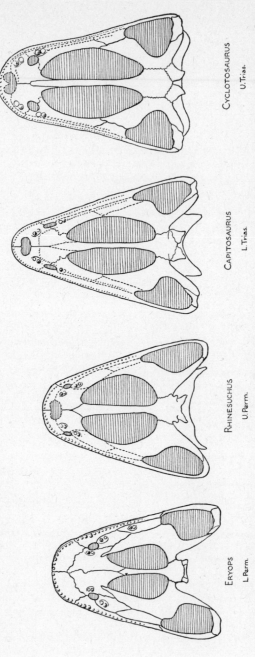

ERYOPS
L.Perm.

RHINESUCHUS
U.Perm.

CAPITOSAURUS
L.Trias.

CYCLOTOSAURUS
U.Trias.

FIGURE 14. A series of drawings of capitosaurine skulls arranged in order of time to show the evolutionary changes in the palate. Note especially the increase in size of the interpterygoid vacuities (cross lined), and the forward movement of the quadrate condyles for the articulation of the lower jaw (until in *Cyclotosaurus* they are on the same plane as the occipital condyles). The gradual change in the character of the neck articulation from a single surface to well-separated condyles is also well shown. *Eryops* after Broom; *Rhinesuchus* after Broom, and Haughton's figures of *R. whaitsi*; *Capitosaurus* founded on Schroeder's *C. helgolandiae*; *Cyclotosaurus* from Fraas' photograph of *C. postumus*.

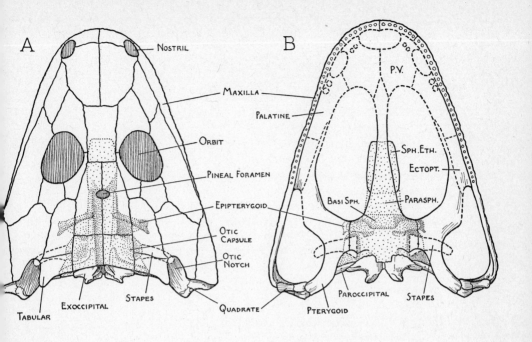

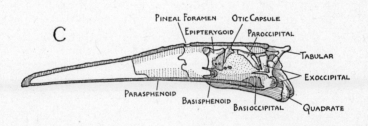

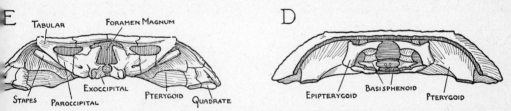

FIGURE 15. The skull of an Upper Permian labyrinthodont. A series of drawings representing the dorsal aspect (A), the palate (B), the occiput (E), a median section (C), and the posterior part of the skull (D) as it appears from in front when the skull is cut through immediately behind the pineal foramen. The brain case in (A) and (B) is shown in stipple as it would appear if the membrane bones of the skull roof and palate were transparent. The interpretation of the reference letters is as follows: *Basi. Sph.*, basisphenoid; *Ectopt.*, ectopterygoid; *P.V.*, prevomer; *Parasph.*, parasphenoid; *Sph. Eth.*, sphenethmoid. This figure is intended to serve as a guide to the structure of the skull in labyrinthodonts in general, and to help the reader to interpret the other figures. All the figures represent a single skull of an animal near *Lydekkerina* from the Upper Permian Lystrosaurus zone of Harrismith, Orange Free State, South Africa. The dermal bones of the skull roof are slightly diagramatized.

43

the general dermal coating. The rigidity of the whole structure is increased by the production of special buttresses which cross from one structure to the other and act as the knees which connect the ribs and thwarts of a boat. The bones involved are constant in number and in their general position.

The teeth which border the mouth of a labyrinthodont are uniform in size and shape. From the mid-line to the nostril they are attached to the premaxilla and are continued backward on the maxilla. Behind the maxilla the lower border of the skull is composed of one or two bones, of which the posterior reaches the extreme posterolateral corner and is called the quadratojugal.

The palate is horizontal, its most anterior portion being supported by a pair of prevomers, flat bones firmly attached to the inner borders of the premaxillae, meeting one another in the mid-line. In many labyrinthodonts a single hole or a pair of holes lies between the premaxilla and the prevomer. These are quite irregular in their occurrence and exist merely to provide head room for enlarged teeth in the front of the lower jaw. Behind the nostril a flat palatine bone, firmly attached to the maxilla, stretches inward as a shelf in the roof of the mouth, its anterior end firmly attached to the posterior border of the prevomer.

The internal or palatal nostril always lies between the maxilla, palatine, and prevomer; it is a small hole lying some distance behind the anterior nostril, so that inspired air must pass through the cavity of the olfactory organ.

Behind the palatine lies a transverse bone, the ectopterygoid. This is attached to the maxilla and to the hinder end of the palatine and has a free posterior border which forms part of the margin of a great hole, the subtemporal fossa. Through this the muscles of mastication, which by their contraction close the mouth, pass downward from the skull to the jaw.

The lower jaw articulates with a special bone, the quadrate, which lies at the posterolateral corner of the skull attached to the inner surface of the quadratojugal. Unlike those we have so far considered, the quadrate is a "cartilage bone" in that it has replaced the posterior end of a cartilaginous bar, the homologue of the upper jaw (palatoquadrate) of a dogfish, which actually existed in the young labyrinthodont as a continuous structure and still survives (except where it is replaced by the quadrate and the epipterygoid) in the full-grown *Eryops*.

Forward from the quadrate, very nearly to its anterior end at the olfactory capsule, the palatoquadrate cartilage is underlain by a great membrane bone, the pterygoid. This is attached to the inner surface of the quadrate and extends upward as a sheet powerfully connected to a great bone (the squamosal) of the skull roof. This vertical quadrate ramus of the pterygoid suddenly ends; the remaining palatal part extends forward in the plane of the palate, forming the outer border of the great interpterygoid vacuity in the bony palate and having its lateral border very firmly attached to the ectopterygoid, palatine, and prevomer.

This series of bones thus forms a rigid structure attaching the quadrate to the upper jaw, capable of giving origin to the muscles which move the lower jaw and of meeting the great stresses produced by their contraction. In a large skull the force exerted by the contraction of these muscles may have been of the order of half a ton.

The structures just described are not in direct contact in the middle line, nor do they provide any attachment to the vertebral column and the body in general. These deficiencies are made up by the attachment of the whole structure to the brain case. As labyrinthodonts are primitive animals their brains are necessarily small, and in accordance with the general rule that large animals have always relatively smaller brains than smaller, closely allied animals, the large creatures with which we are concerned have very small brain cavities and brain cases.

The primitive brain case in embryonic vertebrates surrounds a cavity whose floor is ultimately a continuous mass of cartilage. To this is attached a pair of cartilaginous olfactory capsules anteriorly and a pair of cartilaginous otic or ear capsules posteriorly. The orbit lies between the olfactory and otic capsules, which form part of the side wall of the brain cavity. The remainder of this side wall is a sheet of cartilage, continuous except for certain foramina through which nerves and blood vessels pass into or out from the brain cavity. The roof is usually incomplete, but there is at least a bar joining the otic capsules and a larger stretch of cartilage farther forward.

In well-ossified labyrinthodonts there are three bones in the floor of the brain case, a basioccipital at the hinder end and a basisphenoid in front of it, which lie entirely in the floor and support a pair of exoccipitals surrounding the foramen magnum, through which the spinal cord passes forward to expand into the brain. The otic capsule has two bones, an opisthotic behind and a pro-otic in front, the latter touching the basisphenoid. The summits of the exoccipitals are

connected to one another by a supraoccipital bone. The part of the
brain case lying between the eyes is ossified as a single bone, the
sphenethmoid, which completely surrounds the cavity.

The brain case so formed lies with its dorsal surface in tight contact with the inner surface of the dermal bones of the skull roof, and
is immovably connected to them. The lower and central surface of
the brain case from front to back is underlain by a single membrane
bone, the parasphenoid, which extends forward to the prevomers.
Its lateral edges form the inner borders of the interpterygoid vacuities, and the hinder end is firmly attached to special inwardly directed processes of the pterygoid.

Thus the whole skull is firmly tied together as a mechanically
sound structure and is connected to the body by a contact of the
hinder ends of the basi- and exoccipital (the condyle) with the front
of the vertebral column, and by the attachment of body muscles to
the hinder face of the whole brain case and the membranous bones
which form a frame to it.

The series of palates of capitosaurine skulls shown in Fig. 14 is
arranged in order of time, the individual skulls, which are all about
the same size, being reduced to the same width. It is evident from
this figure that certain changes take place regularly, each skull differing from that which precedes it as its successor differs from it. The
most obvious change is a steady increase in the size of the interpterygoid vacuity, shown by a reduction in the width of the palate
between its lateral borders and the outer surface of the maxilla. This
narrowing is brought about by an eating away of the palatal ramus
of the pterygoid, which in the earliest form meets the prevomer, in
the second fails to establish this contact, in the third just touches
the palatine, and in the last makes very short contact.

A second change vividly shown by this series of drawings is that
at first the quadrate articulation with the lower jaw lies far behind
the occipital condyle, by which the skull is attached to the vertebral column, while by a steady progress of movement forward the
quadrate condyle in the last animal comes to lie on the same plane as
the occipital condyle.

Another change visible on the palate in this series which proceeds
steadily with time is in the mode by which the pterygoid gains its
attachment to the floor of the brain case (see Figs. 14 and 22). In the
Basal Permian *Eryops* the articular process of the pterygoid is short
from front to back and has a rounded lower surface; the bone here
is in fact half of a thick-walled cylinder, the cavity of which ends

by rapidly tapering to an irregular point. This hemicylinder is attached by a jagged and well-interlocked suture to the skin of hard, smooth bone coating the undersurface of a short cylindrical process, the basipterygoid process, which projects from the side of the floor of the brain case, and the underlying parasphenoid. The core of this process is composed of spongy bone which is part of the basisphenoid, and the whole is continued by a cartilaginous extension lying within the recess in the pterygoid.

In the second form, *Rhinesuchus,* the articular part of the pterygoid and the process of the parasphenoid to which it is attached are flat and much longer than in *Eryops.* The nature of the basisphenoid is not known.

In *Capitosaurus* the relation of the pterygoid to the parasphenoid is much as it was in *Rhinesuchus,* but the basisphenoid is very little ossified, the basipterygoid process, which still passes outward to end in a pit in the pterygoid, being entirely cartilaginous. But in *Capitosaurus* a new bone enters into the complex, the exoccipital having extended so far forward that it just touches the pterygoid laterally to its sutural union with the parasphenoid. In *Cyclotosaurus* the sutural junction between the pterygoid and the parasphenoid has become very long, and a new interdigited suture between the exoccipital and the pterygoid has made its appearance as a development of the mere contact which exists in *Capitosaurus.*

The occipital views of the same skulls (Fig. 16) show other important changes. The most obvious of these is that the whole skull becomes flattened, the deep cheeks of *Eryops* passing gradually into the shallow side of *Cyclotosaurus,* the flat upper surface of the skull widening at the same time. This general flattening affects the brain case itself, the distance between its upper and lower surfaces becoming noticeably less. One important result of these changes is to make the axis of the quadrate condyle pass through that of the exoccipital condyles, a condition approached in Fig. 16 but achieved in *C. postumus,* the mechanical significance of which we shall presently investigate.

Another regular change shown in the posterior view of these skulls concerns the occipital condyle. In *Eryops* there is a single condyle much wider than it is high, made of a basioccipital bone in the middle and of the exoccipital at the side. In *Rhinesuchus,* although still single and made by the same bones, the middle portion, which is basioccipital, has become small and is drawing forward. In *Capitosaurus* the functional condyle is double, the basioccipital, though

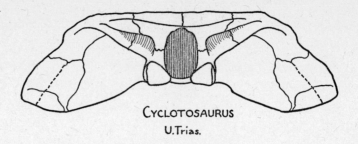

CYCLOTOSAURUS
U.Trias.

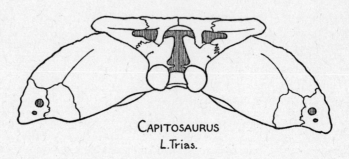

CAPITOSAURUS
L.Trias.

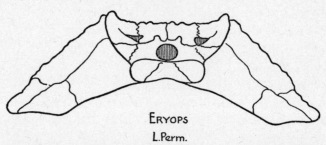

ERYOPS
L.Perm.

FIGURE 16. Drawings of occiputs of three capitosaurine skulls, reduced to the same width for comparison. The evolutionary changes shown are the general flattening, the widening of the "table" (the flat top of the skull), and the development of two occipital condyles. The decreasing ossification of the brain case is shown by the reduction of the basioccipital forming the middle of the single occipital condyle in *Eryops* to a small bone between the condyles of *Capitosaurus*, and its disappearance in *Cyclotosaurus*. Also by the complete covering of the paroccipital, which may be seen in *Eryops*, by outgrowths of the tabular and exoccipital. *Eryops* from the figures of Broom and Case; *Capitosaurus* from a skull from Watford, Cape Province, South Africa; *Cyclotosaurus* from the skull of an early member of the genus from St. Peters, New South Wales, in the British Museum.

48

still of bone, lying so far forward that it takes no part in the support of the head. In *Cyclotosaurus* the basioccipital has vanished entirely, and even the cartilage in which it once developed has become a very thin and feeble structure.

The posterior view of these skulls shows another change, also proceeding regularly with time, which is less easy to see and explain but which is of great importance in any such analysis of the significance of these advances as I propose to make. In *Eryops* the ear capsule is completely ossified, and the paroccipital which forms its hinder surface is widely visible from behind. In order to secure a firm attachment of the dermal skull roof to the brain case, which is in direct contact with it only along a strip around the middle line, two pairs of flanges spread downward onto the occiput from the four bones (the tabulars and postparietals) which form the hinder border of the skull roof. The medial pair are attached to the exoccipitals, and the lateral cover parts of the upper and posterior surfaces of the ends of the paroccipitals. A good deal of the paroccipital thus remains visible from behind. But to a progressively greater extent in later forms the exoccipital spreads outward along the hinder surface of the paroccipital, and the tabular flange spreads downward along that bone until ultimately the two meet and completely hide it from view. At the same time the extent to which the otic capsule is ossified becomes less and less, and its lateral extension toward the tabular is gradually withdrawn. The final result in *Cyclotosaurus* is that the exoccipital and tabular meet in an extensive union and the paroccipital bone vanishes entirely.

APPENDIX

The account of the vertebral column of stegocephalous amphibia given in Chapter II is still, I think, correct; but Romer in various papers, especially in his very useful review of the Labyrinthodonta, has suggested various new interpretations of the conditions actually found. He recognizes the group Labyrinthodonta but includes within it the animals which I have referred to the group Phyllospondyli, i.e., the true branchiosaurs. He regards the embolomerous condition, in which the circumchordal elements are complete perforated discs, as secondary, derived from an earlier rachitomous stage in which the intercentra surround only the lower part of the notochord and the pleurocentra are restricted to the dorsal part of the notochord. Although it is not essential for my purpose in this chapter it may, I think, be useful to comment on these conclusions, which are of considerable general interest.

To deal first with the phyllospondylous vertebrae, the typical phyllospondyl is the animal *Branchiosaurus*, which is known from many hundreds of complete skeletons from the Upper Coal Measures and Lower Permian of Europe. A growth series was described by H. Credner in 1886. The largest individuals in this series have a skull 2.2 cm. in length; in the smallest it is extremely small (less than 1 cm. in total length). Through this growth series the proportions change very markedly, and an intermediate stage between first and last is illustrated (Watson, 1940, p. 225, Fig. 22) whose skull length is about 2 cms. The essential features are the lengthening of the postorbital region and a considerable reduction in the size of the orbit. Between these two stages the internal gill clefts are lost and the gill rakers which bordered them disappear. External gills survive longer than the internal gills and it is not quite certain at what size they cease to be visible, though it would probably be possible now to determine this. It seems to me very improbable that the largest skulls of *Branchiosaurus* could actually have grown into any known rachitomous form; for instance the very short anteroposterior extent of the dermo-supraoccipital and tabulars in them is not paralleled in any rachitomous animal, and the skull roof still lacks the characteristic ornament already well determined in skulls of the same size in the labyrinthodonts *Archegosaurus*, *Actinodon*, and *Benthosuchus*. Furthermore the actual structure of the vertebrae is unlike that of any rachitomous form in that the neural arches, which are the only elements known, extend downward around the notochord for about half its height, while in young individuals of rachitomous form they are restricted to the dorsal surface of the chord. The same ventral extension of the paranotochordal portions of the neural arch is seen in *Miobatrachus*, and in an exaggerated form in frogs, but appears never to have been developed in the true labyrinthodonts. Thus I still regard the branchiosaurs as a genuine group, although I am satisfied that Romer is right in referring the so-called genus *Melanerpeton* (in which, as Credner showed in 1885, a complete, cylindrical circumchordal element may exist) to larval anthracosaurs or perhaps seymouriamorphs. Furthermore it is extremely probable that some other larvae of labyrinthodonts have been referred to the phyllospondyls, a group which after all is of ultimate labyrinthodont origin. Lepospondylous vertebrae are entirely distinct; even in very young individuals these vertebrae are not seen to be composed of individual elements, and even in the smallest examples there is an undivided centrum not yet of its full length, so that intervals occur between successive circumchordal centra. Adelospondylous vertebrae differ from all others in that the centrum is continuous around the notochord and only fuses with the neural arch late in life. The course of evolution of labyrinthodont vertebrae depends ultimately on the time at which the embolomerous, rachitomous, and stereospondylous types first make their appearance. Romer believes that the rachitomous type is the earliest and that the embolomerous, seymouriamorph, and reptilian stages arise from it, as do the neorachitomous and stereospondylous versions (see Romer, 1947, p. 64, Fig. 11).

I would urge that the protorachitomous form, which is his presumed primitive condition, has never in fact been seen; it is purely hypothetical in its occurrence. In the Texan deposits of high Upper Carboniferous and Lower Permian age rachitomous and embolomerous vertebrae are found in a typical form. The only known intermediates between them are those which have been described by Dr. Steen in *Acanthostoma vorax* from the Lower Permian of Niederhässlich, and from high Coal Measure horizons of Linton, Ohio, and the somewhat more ancient Middle Coal Measure horizon of South Joggins, Nova Scotia. It may, I think, be said that typical rachitomous vertebrae do not occur in the British labyrinthodont-yielding horizons, which range in date from some point in the Lower Carboniferous (Visean ?) through the Namurian and up to the extreme summit of the Middle Coal Measures, or even beyond in the zone of *Anthraconauta philipsi*.

These horizons, which are quite numerous, have in the past been ardently explored by very able local collectors, whose collections, in such museums as Newcastle-upon-Tyne and Manchester, contain many thousands of small bones of all kinds. There are, for instance, hundreds of isolated basisphenoids of coelacanths in these two museums. These collections contain from all these horizons very large numbers of typical embolomerous centra and intercentra, but I have never seen any single rachitomous pleurocentrum or intercentrum— bones which, had the loxommids possessed (as Romer suggests) a rachitomous vertebral column, would have been relatively large and abundant, for loxommid skulls are much more abundant than those of the anthracosaurs.

I therefore still maintain the view expressed in Chapter II, holding that the unusual labyrinthodont vertebrae described by Miss Steen in the case of *Dendrerpeton, Dendriasousa, Calligenethlon, Potamo-coston,* and *Acanthostoma,* with *Stegops* and *Erpetosaurus,* represent stages in the development of a rachitomous from an embolomerous type. I propose to return to this subject on some later occasion.

To the series of capitosaurine skeletons may now be added the very remarkable early form *Edops,* described by Romer and Whitter, which is earlier in date and more primitive in structure than *Eryops,* and *Wetlugasaurus* Efremov, of Upper Permian age. The brachiopid series has gained a very late member of Rhaetic age in *Gerrothorax,* and to that group is to be referred the so-called *Acheloma casei* from the Texan Permian, whose skull structure is now known to me and which I hope some day to discuss.

Edops in all respects provides a morphological ancestor to *Eryops;* it has a single, almost circular occipital condyle, its interpterygoid vacuities are small, and the quadrates lie far back. The brain case is remarkably fully ossified, and there is a movable basicranial articulation of the pterygoid. The character of the epipterygoid is what one would have expected in an *Eryops* ancestor, and the general skull-roof pattern is exactly that which might have been anticipated; there is, for example, an intertemporal, lost in *Eryops.*

III. THE MEANING OF STRUCTURAL CHANGE TO THE ANIMAL

In Chapter II I showed how by picking out all the capitosaur-shaped skulls of labyrinthodonts that are known and arranging them in the order of their stratigraphical position it becomes evident that a number of structural changes take place between *Eryops*, the first member of this line, and *Cyclotosaurus*, the last. Although I have drawn and discussed only four successive stages, I am in fact acquainted with some twenty-four species belonging to this series, the majority of which are undescribed or known only by inadequate accounts. The whole of this material reveals no facts inconsistent with the short summary just given, but shows that at any given horizon there is a small range of variation of each character.

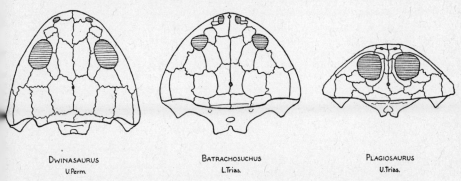

DWINASAURUS
U.Perm.

BATRACHOSUCHUS
L.Trias.

PLAGIOSAURUS
U.Trias.

FIGURE 17. Drawings of the skulls of three brachyopid labyrinthodonts arranged in order of age and reduced to the same width. To show the widening of the head with time and the variation in size of the orbit which makes it improbable that the forms are actual ancestors of one another, although they are clearly close relatives. *Dwinasaurus* after a cast, the photographs published by Amalitzky, and Bystrow's figure; *Batrachosuchus watsoni* from the type specimen; *Plagiosaurus* from Fraas.

It is important for my purpose to build up and examine similar series of skulls of other shapes. The only other completely satisfactory series is that of the brachyopids, which begins with the Upper Permian *Dwinasaurus* from Russia and progresses through the Australian *Bothriceps*, of very similar structure and probable Upper

53

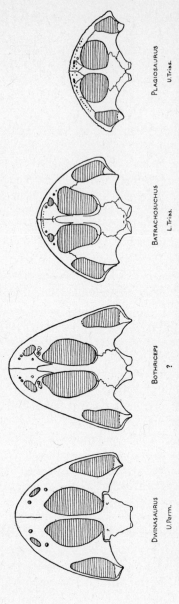

DWINASAURUS
U.Perm.

BOTHRICEPS
?

BATRACHOSUCHUS
L.Trias.

PLAGIOSAURUS
U.Trias.

FIGURE 18. The palates of brachyopids arranged in order of time to show the evolutionary changes. The increase of the interpterygoid vacuities and the changes in the mode of attachment of the pterygoid to the base of the brain case take place here as in capitosaurs, and the forward movement of the quadrate condyle is vividly shown in a comparison of *Bothriceps* and *Batrachosuchus*. *Dwinasaurus* from a cast in the British Museum of Natural History and a figure by Sushkin; *Bothriceps* from the type of *B. australis*; *Batrachosuchus* from the type of *B. watsoni*; *Plagiosaurus* from Fraas and von Huene.

54

Permian age, to the Lower Triassic *Batrachosuchus* of South Africa and the Upper Triassic *Plagiosaurus* of Germany.

Figure 17 shows the dorsal surface of three of these skulls. They widen with time and the orbits become larger, but the pattern of the dermal bones is recognizable throughout.

The palatal views of the same skulls in Fig. 18 are of much greater interest. It is clear that in brachyopids, as in the capitosaurians, the interpterygoid vacuities increase in size, the width of the palate becoming less and the tips of the palatal process of the pterygoid regressing in the same way. In the brachyopids the quadrate condyle always lies anterior to the occipital condyle, but the changes in the attachment of the pterygoids to the base of the brain case follow, even in detail, the same course as in capitosaurs. *Dwinasaurus* agrees with *Eryops*, *Bothriceps* with *Capitosaurus*, and *Batrachosuchus* with *Cyclotosaurus*, while *Plagiosaurus* shows a still further advance in the same direction.

In the occipital view (Fig. 19), although *Plagiosaurus*, the latest, is by far the most depressed form, there is no clear evidence of steady progress because *Batrachosuchus* is much deeper than is the earlier *Dwinasaurus*. But in all other respects the changes, even in this view, are regular and of the same nature as those in capitosaurids. The occipital condyle of *Dwinasaurus* is single, but the basioccipital which forms the middle of it is undergoing reduction. *Bothriceps* agrees exactly with *Capitosaurus*, and *Batrachosuchus* with *Cyclotosaurus*, while *Plagiosaurus* goes still further in the extreme thinness of the cartilaginous floor of the brain case.

In *Dwinasaurus* the flange from the tabular is separated from the exoccipital by a cartilaginous but widely expanded paroccipital; in the later forms the two bones meet in a progressively strengthened suture exactly as in the capitosaurs. In fact the whole course of evolutionary change is identical in the two groups.

The greatest contrast to the brachyopids in skull shape is presented by the long-jawed, fish-eating labyrinthodonts. Of such, three are known: *Archegosaurus* from the Lower Permian of Germany; *Platyops* from the Upper Permian of Russia; and *Aphaneramma* from the Lower Trias of Greenland and the Middle Trias of Spitsbergen. Figs. 20 and 21 show the dorsal and palatal views of these three animals. From them will be seen that the general course of change is in these as it was in the other groups. The interpterygoid vacuities increase in size, the quadrate condyles move forward, the connection between pterygoid and the base of the skull is, in *Archegosaurus*, a movable

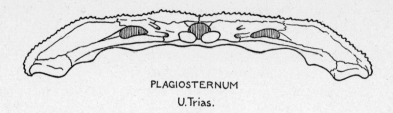

PLAGIOSTERNUM
U. Trias.

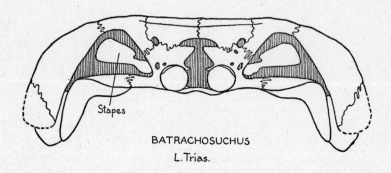

Stapes

BATRACHOSUCHUS
L. Trias.

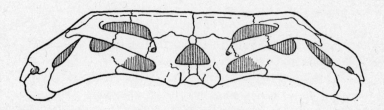

DWINASAURUS
U. Perm.

FIGURE 19. The occiputs of three brachyopid skulls to show the reduction and disappearance of the basioccipital and the conversion of the single condyle of *Dwinasaurus* to the two of later animals. The drawings show also the outgrowth of flanges from the exoccipital and tabular behind the paroccipital, which is visible in *Dwinasaurus*. *Dwinasaurus* from Sushkin; *Batrachosuchus* from Watson; and *Plagiosternum* from Fraas.

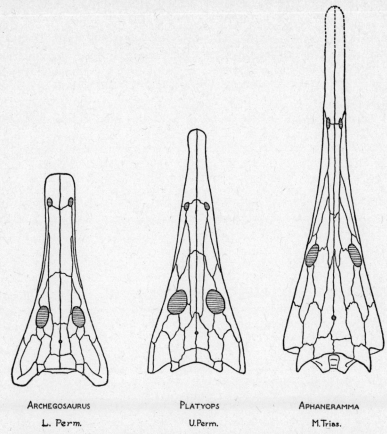

ARCHEGOSAURUS   PLATYOPS   APHANERAMMA

L. Perm.   U. Perm.   M. Trias.

FIGURE 20. Dorsal surfaces of three labyrinthodont skulls of successive ages which have the elongated jaws common to fish-eating vertebrates. They are not closely related. *Archegosaurus* after von Meyer; *Platyops* after Bystrow; *Aphaneramma* after Säve-Söderbergh.

articulation with an ossified basisphenoid, with only a small splint of parasphenoid along the undersurface of the basipterygoid process. In *Platyops* the pterygoid-skull-base articulation is as in *Eryops,* while in *Aphaneramma* there is an extremely long suture between the pterygoid and parasphenoid, and the ventral surface is flat.

The occipital condyle of *Archegosaurus* is exactly like that of *Eryops,* while *Aphaneramma* has two widely separated exoccipital condyles and a reduced basioccipital, like *Capitosaurus. Archegosaurus* has a large paroccipital; *Aphaneramma* agrees with *Capi-*

tosaurus. In fact the course of evolutionary change is exactly what it should be.

This process of comparison can be extended; any series whatsoever of labyrinthodont skulls of similar shape and of Permian and Triassic age will show the same series of changes, although sometimes one of them, usually the depression of the whole skull, may not be exhibited.

This uniformity in the character and direction of the changes

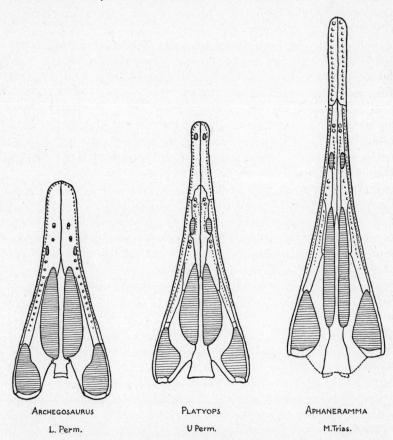

ARCHEGOSAURUS   PLATYOPS   APHANERAMMA

L. Perm.   U Perm.   M.Trias.

FIGURE 21. Palates of the skulls whose dorsal aspects are shown in Fig. 20. Although not a true phylogenetic series these animals, when arranged in their time order, show very well the movement forward of the quadrate condyle and the increased strength of the attachment of the pterygoid to the base of the brain case, which occur in all labyrinthodonts. *Archegosaurus* after Watson; *Platyops* from Bystrow; and *Aphaneramma* from Säve-Söderbergh.

of structure which take place with time in the skulls of labyrinthodonts can be seen very clearly by comparing the whole group of Lower Permian genera with all those of Triassic age. Such a comparison, moreover, is instructive because it shows that the rate at which each of these evolutionary advances takes place may differ considerably in what are obviously independent, though allied, stocks. Thus the whole character of the base of the brain case in the Lower Triassic brachyopid *Batrachosuchus* is identical, not with that of its contemporary *Capitosaurus* but with the Upper Triassic successor of that genus, *Cyclotosaurus.*

The skulls of the Lower Permian genus *Trimerorhachis* are as flattened as those of any Upper Triassic labyrinthodont, but in compensation the occipital condyle of that animal is much more primitive than that of its contemporary *Eryops.* This principal of "compensation," whereby an exceptionally advanced structure of some part of an animal is balanced by the retardation of the evolution of some other part, was first enunciated by the late Dr. W. D. Matthew and seems to have a very wide application. It can be illustrated among fossil mammals or echinoids as well as in many reptile histories. It is perhaps of great theoretical importance, and will be discussed in Chapter VIII.

It is now necessary for the further development of my argument to return to the consideration of the evolution of the various groups into which the labyrinthodonts can be divided on the basis of the shape of their heads. The fact that all these amphibia pursue similar courses of evolutionary change can be accounted for in two ways only. Either we are viewing the history of a group in which there was a single stock whose character was changing steadily, from which at each of many horizons arose side branches with different head shapes fitting them for diverse modes of life, these adaptive twigs having necessarily very similar characters at any time. Or we have a group in which many basal stocks, of different habits, pursue parallel evolutionary courses. There is no doubt that in the main the second alternative is true.

The only way in which we can discriminate between the two views is by obtaining evidence of the persistence in all the members of one series of a definite structure or structures, not found in the members of any other series and not to be accounted for on any ground of adaptation to a special mode of life, which shows that they are closely related to one another. If it should be impossible to discover a structure which does not change during the history of the series, then we must fall back on one which exhibits evolutionary change but which is nonetheless restricted in its occurrences to a single series of animals. The capitosaurine skulls show such a character (Fig. 22). In *Eryops,* behind the basipterygoid processes the lateral margins of the parasphenoid are turned up dorsally so that in the bony skull they form the lower border of the hole in the outer side of the ear capsule, the fenestra ovalis, into which fits the footplate of the stapes, a bone whose function it is to transmit sound waves from the air to the organ of hearing. A special process of the stapes is attached to this free border of the parasphenoid. Behind this opening the parasphenoid underlies the anterior corner of the exoccipital, being there raised into a small eminence to which certain muscles are attached. The palatal surface of the bone shows a pair of depressions bounded by two ridges. In succeeding members of the series, as the exoccipital grows farther forward the free margin of the parasphenoid, between it and the attachment of the pterygoid, becomes progressively shorter, until in *Rhinesuchus* it is almost entirely represented by a small, thick process to which the stapes is still attached. The muscular insertions on the back of the parasphenoid move forward at the same time, and the pair of shallow depressions which lie before them in *Eryops* become deeper and are indeed overhung by a strong ridge. In the later *Laccosaurus* the change proceeds in the same direction, the process of the parasphenoid to which the stapes is attached becoming reduced to a small point. In *Capitosaurus* this point remains as a rudiment, no longer touching the stapes, and the ridges which limit the depression on the lower surface of the parasphenoid now lie very close to one another. In the small and early *Cyclotosaurus, C. stantonensis,* the same ridges are still to be seen, but in the later *C. posthumus* only one survives, and that is a mere thin line.

I know no labyrinthodont outside the capitosaur series in which these structures of the base of the skull are to be found, and regard them as evidence that the members of this series are actually blood relations; that in fact we are dealing with a real genetic series and not

FIGURE 22. The floor of the brain case, with the attached pterygoids, of a series of capitosaurine skulls arranged in order of time. This series shows that small characters of the brain case change by such small steps that they can be traced throughout the series, and give evidence that it is a genuine or very near approach to a true phyletic line. The change in the character of the attachment of the pterygoid to the basis cranii and of the occipital condyle is well shown. *Eryops,* original; *Rhinesuchus* from F. R. Parrington's skull of *R. Nyasensis; Laccosaurus* after Haughton; *Capitosaurus,* original from a skull from Cape Province, South Africa.

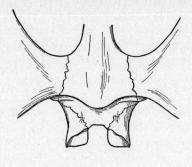

CAPITOSAURUS
L. Trias.

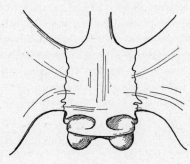

LACCOSAURUS
U. Perm.

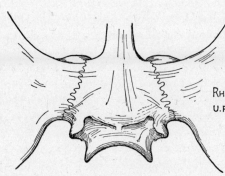

RHINESUCHUS
U. Perm.

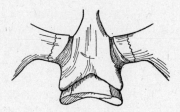

ERYOPS
L. Perm.

61

with an artificial assemblage of animals which happen to illustrate the stages of a change in structure.

The brachyopid skulls described in the beginning of this chapter all possess certain characters which are found neither singly nor together in any other animal. In all of them the squamosal wraps around the back of the quadrate and then turns backward as a ridge which lies parallel to a similar posterior border of the pterygoid. In all of them the pterygoid is turned downward on the inner side of the subtemporal fossa for the masticatory muscles, so that the palate is a broad U-shaped trough. In all of them the occipital condyle lies a long distance behind the posterior border of the dermal roof of the skull, so that the occiput slopes sharply forward. Thus the brachyopids form another real genetic series, shorter than that of the capitosaurines but showing that the two pursued identical evolutionary courses.

When, however, we turn to the long-snouted forms, *Archegosaurus*, *Platyops*, and *Aphaneramma*, we find it impossible to discover any characters, other than gross shape, which, while common to all these, are not also shared by other labyrinthodonts. It is thus in the highest degree improbable that these three forms are really closely related. The series is artificial, illustrating the general morphological changes common to all labyrinthodonts but not showing the history of a group.

It is fortunately possible to go beyond this purely negative conclusion. *Aphaneramma*, which is found in the Marine Lower Trias of Greenland and in the Marine Middle Trias of Spitsbergen, is the most elongated member of a group of very closely allied animals whose members are found in association with it in Greenland and Spitsbergen, and independently in the Marine Lower Trias of India and the Lower Triassic fresh-water deposits in Germany and South Africa. These animals form the family Trematosauridae. That they are indeed very close relatives is shown conclusively by the occurrence in all of them of an extremely striking and unique structure in the base of the skull. Here the pterygoid meets the parasphenoid in a long suture, the two bones growing backward so as almost completely to hide the exoccipital in a palatal view and to form a shelf of bone below the tympanic cavity, the air-filled drum of the middle ear.

All trematosaurids have a skull which is triangular in plan, the lateral margins formed by the maxillae being essentially straight and approaching one another so that the rounded anterior extremity of the snout is very narrow. They differ only in proportion, essentially in the angle that the sides make to one another, and taken together the species completely cover the range between those contemporaries, *Lyrocephalus,* in which the width is about 100 per cent of the total length, and *Aphaneramma,* in which it is only 27.5 per cent.

The simplest explanation of their occurrence is to suppose that they represent the results of a very rapid adaptive radiation whereby an animal of the general proportions of *Lyrocephalus* gave rise to the extremely long-headed forms which analogy with fish, crocodiles, and whales shows to have been fish eaters. This view is consonant with the fact that in all those characters which change with time in labyrinthodonts in general the trematosaurs are identical. There is one obvious exception to this statement: the narrow-skulled *Aphaneramma* is clearly much deeper for its width than the wide *Lyrocephalus.* If, however, we compare drawings of the occiputs of these two animals (Fig. 23), we find that the posterior surfaces of their

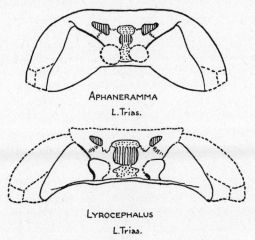

APHANERAMMA

L. Trias.

LYROCEPHALUS

L. Trias.

FIGURE 23. The occiputs of the contemporary labyrinthodonts *Aphaneramma* and *Lyrocephalus,* to show the deepened skull associated with the very elongated snout of the former animal. This deepening is an adaptation cutting across the evolutionary trend. *Aphaneramma,* redrawn from Säve-Söderbergh; *Lyrocephalus,* original.

brain cases are generally similar not only in structure but in proportion, and that the differences lie in those lateral parts of the skull which include the quadrates and in the cavities within which the masticatory muscles lay. This region is both narrower and deeper in the long than in the short head, clearly to subserve mechanical requirements.

It follows from this that while in many cases labyrinthodont stocks which separated from one another in early times subsequently pursued parallel evolutionary lines, it did sometimes happen that from some member of one or the other of these stocks there arose a side line, often short-lived, which without any advance in structure branched out rapidly and gave origin to animals fitted for some special mode of life.

I have examined all the labyrinthodonts known, usually from original materials but sometimes only from the published descriptions, to see how far the great variety of forms that existed can be accounted for as the results of rapid and repeated adaptive radiations from the successive members of one or more relatively unspecialized, fundamental, slowly and regularly evolving stocks. Consideration of such details as those I have used to show that the capitosaurine and brachyopid series are genuine phylogenies indicates the improbability that any such explanation can have general applicability. This view is supported by a consideration of the differences of relative rates of advance of those parts of the skulls which change in the same direction with time in all labyrinthodonts.

We are thus driven to the conclusion that this group consisted of a considerable number of stocks which, from the time of their separation, independently pursued similar evolutionary courses for those parts of their skeleton which formed the subject of the preceding analysis.

It is therefore legitimate and convenient to talk of evolutionary *trends* common to all labyrinthodonts, and to grade them on the basis of the *advance* which they have made in following out these trends in all parts of their structure.

The fact that a bulk comparison of all the Lower Permian with all the Triassic labyrinthodonts shows that evolution within the group proceeds in accordance with the same series of trends, whatever the sizes and shapes of the animals, makes it necessary to consider the habits and mode of life of some members of this group before it is possible to investigate further the significance of these trends.

In Figs. 24 and 25 I have placed together drawings of the skeletons of *Eryops* and *Cyclotosaurus*. *Eryops* is about five feet in total length, *Cyclotosaurus* some ten. It is evident from these drawings that *Eryops* is a round-bodied animal whose head is wedge shaped because the muzzle is flattened, while the postorbital part of the skull and lower jaws taken together are nearly as deep as they are wide. The shoulder girdle is a very large, massive, and rigid structure tied down to specially widened ribs by a powerful musculature. The pelvis is equally solid and well attached to the sacrum, and remarkably narrow. The size and character of these limb girdles, taken together with the character of the ribs, show that the back was arched and the whole

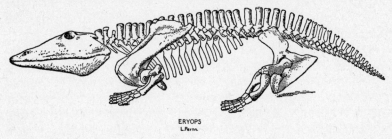

ERYOPS
L.Perra.

FIGURE 24. Restoration of the skeleton of *Eryops* from the Lower Permian of Texas. To show the flattened head, round trunk, and heavy legs of this essentially terrestrial animal. The pelvis is not quite correctly articulated. After Case, 1911.

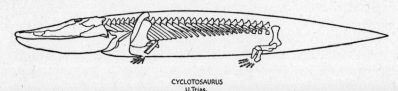

CYCLOTOSAURUS
U.Trias.

FIGURE 25. Restoration of the skeleton of *Cyclotosaurus*. To show the flattening of the whole of the animal in front of the hind leg, and the extremely short and feeble limbs. Total length nearly 3 meters. From a skeleton from the Upper Trias of St. Peters, New South Wales, in the British Museum. The tail is preserved but has not yet been prepared; the length shown is correct.

trunk, perhaps, rather deeper than it was wide. The limbs are those of a heavy-bodied land animal, the feet, though very short, being spreading and having a large surface area. The limb bones are fully ossified, fitting together so accurately that the possible directions and extent of their movements on one another can easily be determined by trial; even the small bones of the wrist and ankle are perfectly finished. There is in fact no doubt whatever that *Eryops*, at any rate when full grown, was a terrestrial animal living most of its time on land, returning to the water perhaps only to breed.

Cyclotosaurus can be compared bone by bone with *Eryops*.

Throughout it is evident that all the structures are morphologically similar, the only real differences being in proportions; but the whole animal is very different. The flattening which was to be seen in the snout of *Eryops* has spread backward over the whole body, except the tail. The skull and lower jaw are of such a shape as to show that the width of the neck was nearly twice its height, and the shoulder girdle shows that this proportion continues down the body. The height of the pelvis fixes very nearly that of the body at the hind legs and shows that even there the body was flattened, which is confirmed by the straightness of the ribs. The great breadth of the body causes the forelegs to be absurdly separated from one another; and they are extremely small, about the same size as in an *Eryops* of half the length and perhaps one sixth of the weight. The individual limb bones are always incompletely ossified, the articular surface remaining cartilaginous, and there are no bones in wrist or ankle.

It seems quite impossible that *Cyclotosaurus* could ever have lived on land; it must throughout its life have rested on the bottom of a pond or river, lying unmoving for long periods, making snaps with its huge jaws at passing fish or reptiles on those rare occasions when it was hungry. It lived, in fact, the life of a Japanese giant salamander, which externally it may have greatly resembled.

There is one piece of nearly direct evidence to show that *Eryops* was a land- and *Cyclotosaurus* a water-living animal. All fish, and the thoroughly aquatic living amphibia, possess a sense subserved by a structure called the lateral-line system, which functions only when immersed. The system in primitive fish is housed in a series of canals which have a very characteristic and uniform arrangement and which penetrate certain of the dermal bones of the skull. In primitive fossil amphibia these canals lie in grooves in the bones. In *Eryops,* although these grooves seem to occur in young and middle-sized individuals, they vanish in the full-grown adult. In *Cyclotosaurus,* however, they are to be seen in specimens of any size and become relatively enormous furrows in the really large skulls. It therefore appears that capitosaurs show the transformation of land-living animals into predominantly aquatic descendents.

We may now discuss how far the changes in the structure of labyrinthodonts, which proceed steadily with time in series in accordance with the evolutionary trends proper to the group, are interrelated. That is, we must endeavor to discover whether some or all of the actual, observed advances may not be merely the necessary sequels of a single or of very few fundamental causes. The only method of attacking this problem is to determine the mechanical effects of the changes which take place.

It is convenient to begin with a part of the animal we have not yet considered, the lower jaw. In all labyrinthodonts the lower jaw is a complex structure built up by the close attachment to one another of a number (usually seven) of membrane bones. These are so arranged that they form a tubular girder, on whose upper surface the long, close-set series of teeth is placed, the anterior end of the jaw sometimes supporting one or two large tusks placed further within the mouth. The muscles whereby the mouth is closed passed downward from the skull through a large hole in the upper surface of the hinder part of the jaw and were actually inserted in its interior, in part on the membrane bones and in part on the remains of the original embryonic cartilage jaw. The lower jaw is articulated with the skull by a cylindroid hollow on its upper surface at the hinder end. This glenoid cavity lies on a bone which replaces part of the original cartilage but which is always continuous with the uppermost of the dermal bones of the jaw. The jaw is long, slender anteriorly, and subjected to great stresses because the muscles which move it must, in a large skull like *Cyclotosaurus*, be capable of exerting an immense pull, something of the order of half a ton. In order to save material and weight, the lower jaw must be designed on sound mechanical principles; that is, it must be deepest at the point where the pull is applied, and taper from there to the anterior end; it should

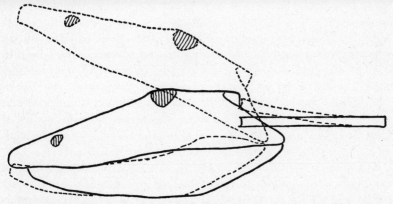

FIGURE 26. Outline of the skull and vertebral column of *Eryops*. To show that if the lower jaw rested on the ground the mouth could be opened only by raising the head, and that such a movement would involve raising also the anterior part of the body. The raised position is shown in broken line.

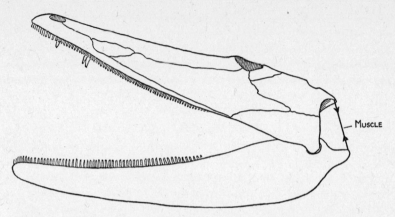

— MUSCLE

FIGURE 27. Outline of the skull and lower jaw of *Cyclotosaurus*. To show
how the mouth is opened by raising the skull through contraction of the
musculus depressor mandibuli, which is attached to the tip of the retroarticular process and labeled "muscle." Such movement involves no displacement of body. From a skeleton of *Cyclotosaurus* from the Upper Trias
of St. Peters, New South Wales, in the British Museum.

have the shape of a cantilever. In *Eryops* it does in fact conform
very closely to this expected shape.

In the land-living *Eryops,* where the head is carried clear of the
ground, the mouth can be opened by dropping the lower jaw, which
is pulled backward and downward by muscles inserted on the lower
surface, and the angle to which it can be opened can be increased by
raising the head a little by contraction of neck muscles. In *Cyclotosaurus,* however, which is so flat that the lower jaw rests on the ground
when the animal is inactive, and in which the front legs are so short
and feeble that the animal cannot have raised its head as a whole, some
other method of opening the mouth is necessary. As in the crocodile
the only way this can be done is by raising the skull. Since the skull
is extremely low and the occipital surface shallow the neck muscles
which raise the skull work at a very great mechanical disadvantage,
and it is clearly desirable to supplement them by others. This was
done in *Cyclotosaurus* as follows. The lower jaw is carried out behind its articulation with the skull by a short, powerful retroarticular
process, and a great muscle, the musculus depressor mandibuli, attached to its end passes upward to be inserted onto the posterior
edge of the dermal skull roof at the highest possible point (see
Fig. 27). Contraction of this muscle will raise the skull if the lower
jaw rests on the ground.

We should thus expect to find that progressive flattening of the head in capitosaurs was associated with a steady increase in size of the retroarticular part of the mandible, and our expectation is justified (see Fig. 28). Indeed all flat-headed labyrinthodonts show some retroarticular process, and in all those of the Upper Trias it is ex-

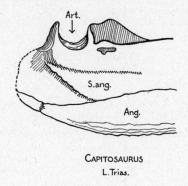

CAPITOSAURUS
L. Trias.

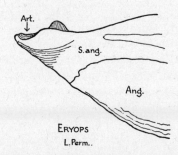

ERYOPS
L. Perm..

FIGURE 28. The posterior ends of the right lower jaws of *Eryops* and *Capitosaurus*, to show the development of a special retroarticular process behind the articulation (*Art.*) in the later forms. *Ang.*, angular; *Art.*, articular facet; *S. Ang.*, surangular. *Eryops* after Broom; *Capitosaurus* from a South African specimen.

ceedingly well developed. Thus the development of this process is a necessary corollary of the flattening of the head and may have no independent cause.

It is plausible to relate another feature in skull structure to the increasing difficulty in raising the head which results from the flattening; this is the widening of the occiput which leads to an increase in the area of the flat upper surface of the skull and to a more vertical position of the cheek (see Fig. 29). This change of shape gives a larger area for the attachment of the dorsal neck muscles and enables the musculus depressor mandibuli to be placed more nearly vertical so that its pull on the skull and jaw is more favorably directed. Thus this trend of change is also a mere sequel of the general flattening.

One of the most easily seen and striking of all the evolutionary changes in these skulls is the steady movement forward and upward of the articular surface of the quadrate bone, to which the lower jaw is articulated. In *Eryops* this joint lies far behind and below the occipital condyle (see Figs. 13 and 14), in a position inherited from its ancestors, which is retained because it is mechanically favorable in a round-bodied animal whose head is carried clear of the ground, which needs a very long jaw and has a very small brain and brain case.

If, however, we imagine that a flattening of the body was carried as far as it is in *Cyclotosaurus* in an animal in which the lower-jaw articulation and the occipital condyle remained in the same relative position as they occupy in *Eryops,* we can very readily see that the animal considered as a structure would be extremely ill designed. The lower jaw would necessarily rest on the ground, and the mouth would be opened by raising the head. But because the occipital condyle would lie far in front of the jaw articulation, such a movement would involve raising the whole of the anterior part of the body from the ground and sliding the lower jaw forward, scraping the lower surface of the head (see Fig. 26).

The conditions which exist in late *Cyclotosaurus,* where the occipital condyle and the quadrate articular surfaces are solids of revolution about the same axis, completely solve these difficulties; the movement of the head involves no displacement of the lower jaw and anterior part of the body. This result is secured by the slow and steady migration forward of the quadrate, which is observed not only in the capitosaur but in most of the independent stocks of the labyrinthodonts.

Consideration of the general body shape of *Cyclotosaurus,* of its extreme width in proportion to its height, and of the way in which the huge shoulder girdle is packed away immediately behind the head, shows that there can be no possibility of any lateral movement of the skull. It could rotate on its condyles up and down, but had no other freedom. This restriction is reflected in the nature of the joint between the skull and the vertebral column. The two exoccipital

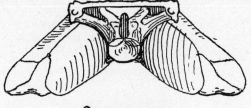

ORTHOSAURUS
U.Carb.

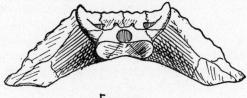

ERYOPS
L. Perm.

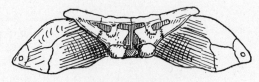

CAPITOSAURUS
L.Trias.

CYCLOTOSAURUS
U.Trias.

FIGURE 29. The occiputs of four labyrinthodonts reduced to the same width and arranged in order of time, to show the persistence of the trend which leads to a depression of the head. All after Watson.

condyles are well separated from one another; their articular surfaces form parts of the same solid of revolution developed on the same transverse axis. They are so extensive as to show that the head could be raised through a considerable angle and they are di-

rected somewhat downward to fit into corresponding somewhat upwardly directed concavities on the atlas and thus reduce the liability
to dislocation and the strain to which the neck muscles would be subjected in keeping the head and neck in articulation.

The nature of the single condyle of *Eryops* is such as to allow of
little movement of any sort, and it seems evident that if from an
early stage in the evolution of the capitosaurine line dorsoventral
movements of the head became customary and lateral movements
became impossible, the changes we have observed in the nature of
the articulation of the skull with the neck might on mechanical
grounds have been expected.

It is possible and necessary to continue much further this analysis
of the labyrinthodont skull, considered as an engineering structure,
designed to be subjected to definite forces. In *Eryops* the articulation of the lower jaw is with the posterior end of a quadrate bone
firmly held between the squamosal and quadratojugal, which are
part of the functionally continuous, thin dermal skull roof, and the
quadrate ramus of the pterygoid, a deep sheet of bone standing
nearly vertical. The effect of the stress to which this region is subjected by the contraction of the jaw muscles is to force the articulation upward, a movement which if allowed to take place would distort the skull by bending inward the curved lateral border of the skull
roof forming the margin of the mouth, and by driving the point at
which the pterygoid is attached to the base of the brain case upward.
The skull is strengthened to prevent its deformation by strutting out
the outer margin with that part of the palatal ramus of the pterygoid
which is attached to the ectopterygoid, and by the solidity of the
well-ossified brain case itself. Furthermore the pterygoid is converted
into a cantilever (analogous to the cantilever spring of a car) by
its attachment, essentially a pivot, to the basipterygoid process, and
by the carrying of its palatal ramus to a firm anchorage far forward
in the skull, where it has a sutural attachment to the prevomer. This
part of the skull is thus well designed to withstand the stresses to
which it is subjected. It is, however, obvious that the function of the
pterygoid palatal ramus as one arm of a cantilever would best be
served if it ran straight forward from the 'articulation to the brain
case; and it is desirable to see if some explanation of its actual shape,
bowed outward to form the lateral border of the interpterygoid
vacuity, cannot be found.

It must always be borne in mind when making a mechanical
analysis of any part of an animal that the latter is not a free agent.

Its adult structure is necessarily based on the materials available as an inheritance from its forebears, and these are limited because they must arise during the embryonic development of the animal from pre-existing structures, which may have impressed on them special shapes and relationships because they subserve definite functions in the developing but living foetus. Dr. Säve-Söderbergh has recently shown in trematosaurids, and I have long known in *Eryops*, that the pterygoid in the adult skull retained its old relationship to the unossified palatoquadrate cartilage, which is the embryonic upper jaw. In the adult skull this cartilage was a narrow and thin strip lying on the upper surface of the palatal ramus of the pterygoid throughout its length, until, in fact, as Säve-Söderbergh has shown, it came into contact with the lower surface of the olfactory capsule. There is no doubt that ossification of the pterygoid first began in direct association with the cartilage in labyrinthodonts, as it does in frogs. As the palatoquadrate is the only skeletal structure which can support the lateral portions of a wide head in an embryonic amphibian, it seems inevitable that it should be bowed outward as the skull becomes broader and takes on the form actually retained in adult specimens of *Eryops*. Thus the interpterygoid vacuity in labyrinthodonts may arise merely because the pterygoid failed to extend inward over it to form a thin sheet of bone with no important function, which was inconvenient in another way.

Even in *Eryops* the distance between the roof of the skull and the palatal surface of the parasphenoid in the neighborhood of the orbit is so small that it is doubtful if a spherical eye, of size appropriate to the orbit, could be contained in it. The eyeball must in part have projected above the general level of the head as it does in frogs. So placed it was exposed to risk of serious damage, especially if it rested on a solid bony floor to the orbit. The existence of the interpterygoid vacuity below the orbit enables the eye to be depressed as it can be in frogs, the soft palate bulging out below it so that it is not subjected to great pressure by any accidental contact.

The skull of *Cyclotosaurus* differs from that of *Eryops* in that the articular surface of the quadrate is placed so far forward that it lies only slightly behind the attachment of the pterygoid to the brain case. This disposition of the parts greatly alters the direction and magnitude of the stresses to which the hinder parts of the skull are subjected. The action of the biting muscles still forces the quadrate upward, but the stress now tends to rotate the pterygoid as a whole on the basipterygoid process only to a very small degree, and it is no longer necessary to make an extremely forward anchorage for the palatal ramus of the pterygoid in order that it may act as a cantilever arm to prevent such rotation. The palatal ramus is thus both shortened and weakened, a process carried out by reducing the extent to which it grows in below the interpterygoid vacuity, which hence automatically increases in size by the eating away of its lateral border. But the alteration of the position of the quadrate renders the forces tending to distort the dermal skull roof even greater; the effect of the action of the jaw muscles, if uncontrolled, would be to force the margins of the skull outward. This danger can be met by retaining the original connection made by the pterygoid between the ectopterygoid and the base of the brain case but converting it from a compression to a tension member. A pull of this kind can be transmitted safely from one membrane bone to another only through a deeply interlocked suture. This, then, is a possible explanation of the powerful connection between the pterygoid and ectopterygoid which survives even in the most advanced skulls, and of the huge exaggeration of the suture between the pterygoid and the parasphenoid which appears in them.

Despite these provisions the adequate support of the quadrate seems to have been difficult and another bone, the epipterygoid, which in the earliest form known has no obvious mechanical significance, becomes so changed that it forms two struts directly connecting the quadrate ramus of the pterygoid to the upper part of the brain case. In *Eryops* this bone rests on the outer side of the pterygoid, so that it stretches up the anterior edge of the vertical quadrate ramus. The base of the bone is in contact with the outer end of the basisphenoidal (cartilage bone) part of the basipterygoid process. The anterior end of this bone faces the groove on the upper surface of the pterygoid which holds the thin strip of palatoquadrate cartilage. From the free anterior and upper border of the bone a stout process, the processus ascendens, projects upward and a little inward. In some specimens, but not in all, this process touches the top of the side wall of the brain case at a point where it is so little ossified as to be incapable of taking much pressure. Otherwise, at any rate in most specimens, neither the epipterygoid nor even the upper corner of the pterygoid touches any part of this brain case.

In such later capitosaurids as *Laccocephalus* the processus ascendens has become a thick, deep, transversely placed sheet of bone, still articulating with an unossified part of the brain case but doing

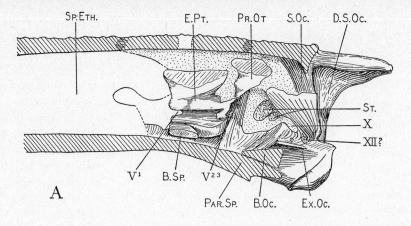

A

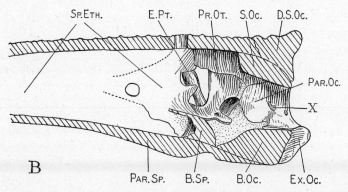

B

FIGURE 30. The inner surface of the right half of the brain case of A, *Capito-saurus*; B, *Eryops*. To show the much less extensive development of bone in the former Lower Triassic animal than in the latter from the Lower Permian. Reference letters: *B.Oc.*, basioccipital; *B.Sp.*, basisphenoid; *D.S.Oc.* dermo-supraoccipital; *E.Pt.*, epipterygoid; *Ex.Oc.*, exoccipital; *Par.Sp.*, parasphenoid; *Pr.Ot.*, pro-otic; *S.Oc.*, supraoccipital; *Sp.Eth.*, sphenethmoid; V^1, V^{2,3}, X, XII, the foramina for these cranial nerves. This figure should be used in comparison with Figs. 22 and 15. After Watson.

so over a very large area. The epipterygoid still fails to meet any part of the otic capsule, but the upper corner of the pterygoid has turned inward so that it has gained a sutural attachment to the upper end of the anterior part of the capsule, the pro-otic; and this corner is the only ossified region of the ear capsule. In *Capitosaurus* the process goes further and the attachment of the pterygoid to the pro-otic is supplemented by the upgrowth of a special otic process from the epipterygoid which abuts firmly against the front face of the pro-otic. By this series of changes the epipterygoid comes to have a "three-point suspension" on the side of the brain case and is in a position to afford very powerful support through the pterygoid to the quadrate.

Thus very many of these evolutionary changes which take place in accordance with trends common to all labyrinthodonts can be shown to be sequels of one process, the steady depression which besets the whole anterior region of the body. The rule of "parsimony of hypotheses," Ockham's razor, suggests that we shall be wise to regard the single phenomenon which is responsible for those more obvious changes as that whose solution has yet to be sought.

But there are other trends, in accordance with which the evolution of labyrinthodonts proceeds, which cannot be shown to be causally connected with the general flattening. The most important of them is the downgrowth, from the tabulars and post parietal bones forming the hinder border of the skull roof, of special flanges on the occiput which rest on the posterior face of the otic capsules and ultimately, partly by their own increase in length and partly by an extension of the exoccipital, come to be firmly united to that bone. A very short inspection of actual skulls shows that this trend is intimately connected with the general decrease in ossification of the brain case. This decrease seems to be brought about by a delay in the beginning of the process. It is first visible in *Eryops* where a mass of the bone of the brain case between the basioccipital and the basisphenoid remains without bony replacement, and the extreme tip of the basipterygoid process fails to ossify even in large individuals. In *Capitosaurus* there is very little bone in the ear capsules and none above the brain, while both basioccipital and basisphenoid remain very small bones even in very large skulls. In *Cyclotosaurus* the exoccipitals, which have spread essentially as membrane bones around the ear capsules, are the only bones to be found in the whole brain case. In *Metoposaurus* of the Upper Trias, contemporary of *Cyclotosaurus*, even the quadrate bone does not become bony until the animal has grown to a skull length of about twelve inches, two-thirds of the adult size.

This delay in the onset of ossification applies to cartilage bones throughout the skeleton, and is a very curious and characteristic phenomenon. The dermal bones, even of the animals which exhibit it in the highest degree, are extraordinarily massive; the interclavicle of *Cyclotosaurus* for example is about a foot square and three-quarters of an inch thick. There was hence evidently no lack of calcium and phosphate in the blood plasma. It is evident that the preparatory changes in the cartilage which initiate the first laying down of bone in the perichondrium do not take place, or rather that they do so very late in life. That this is indeed the case is shown by the fact, which may be observed in most suitably preserved Triassic labyrinthodonts, that the perichondral ossification usually extends considerably beyond the area of true endochondral bone.

This delay in the onset of the appearance of cartilage bone in animals which possess powerful membrane bones is exhibited in many different groups of vertebrates. Stensiö has shown that it occurred in cephalaspids and coelacanth fish. It has long been known in Dipnoi, and recent work makes it clear that the sturgeons have the same history, and that to a less extent the osteolepid fish are less well ossified in Coal Measure times than they were in the Middle Old Red Sandstone.

The existence of these widespread parallels to the conditions in labyrinthodonts shows, I think conclusively, that the phenomenon of delayed ossification in cartilage has nothing to do with the flattening of the head, which is the basal cause of so many of the evolutionary trends in these animals.

There are, of course, other changes in the structure of the group which are not related to either cause. For example, the occipital region is progressively shortened, so that the hypoglossal nerve which primitively passes out through a foramen on the exoccipital comes to lie entirely behind the skull. The primitively five-fingered hand loses one digit; and the pubis, well ossified in all Lower Permian species, ceases to become bony, although the ischium retains its full development.

APPENDIX

In 1938 *Dwinasaurus* was described very well by Bystrow who confirms the opinion I had earlier formed that it is a neotenic form with persistent external gills. It may therefore be expected to retain some larval features, even in its skull, and thus to provide a less satisfactory term in the series than it otherwise should do. Nonetheless it seems to fall quite well into place (Bystrow, 1935). The only important feature which needs to be referred to is that the articulation between the basis cranii and the pterygoid is a movable one, though its nature is not, I think, entirely clear. The long-jawed labyrinthodonts *Archegosaurus, Platyops,* and *Aphaneramma* have recently been redescribed; *Platyops* by Efremov (1933) in the first instance, and Bystrow, and *Lonchorhynchus* by Säve-Söderbergh. The relations of these animals to one another have been discussed by Säve-Söderbergh, Bystrow, Efremov, and Romer. Most of these authors are prepared to find a true (though approximate) phylogenetic series in them. I have examined *Platyops* in Moscow and find it to be, as Efremov has shown, remarkable from the extreme dorsal position of the articulation between the lower jaw and the quadrates. The material is very well preserved and the facts are certain. This condition is not, so far as I can see, foreshadowed in *Archegosaurus;* nor can I see any special resemblances between the two skulls and that of *Aphaneramma,* except those dependent on a great prolongation of the preorbital part of the skull.

When we pass on to *Aphaneramma* or *Lonchorhynchus* we find that in dorsal view there is a genuine similarity with *Platyops,* but we are faced with the position, which Romer has recently pointed out, that *Aphaneramma* is an extreme member of the group of which *Trematosaurus* is the longest known, and these animals are contemporary and most of them marine. It seems, therefore, evident that the group as a whole cannot have been derived from the extremely long-headed *Platyops,* and that it represents an example of the rapid adaptive radiation of a single type of more normal skull proportions of a kind whose existence I anticipated in my 1919 paper.

IV. ORIGINS

THE LAST CHAPTER showed that there is a great series of evolutionary changes of structure, which proceed regularly with time and are common to all the varied stocks of the labyrinthodonts, and that this steady progress is uninfluenced by the adaptations and mode of life of the animals considered; it continues uninterruptedly as the animals change from a terrestrial life to one entirely in water.

If we assume, as the evidence justifies us in doing, that the trends of these changes be constant, by projecting them backward we should be able to predict the structure of the Carboniferous ancestors of the Lower Permian creatures which formed the starting point of our analysis.

The ancestor of *Eryops* is likely to have had the same general skull contour, but its head and body should have been so high that the neck was circular in transverse section. If this were the case then the whole series of structural changes which are causally connected with the flattening would be absent. The lower jaws would have no retroarticular projection, because the mouth could be opened in the ordinary way without raising the head. The quadrate articulation would lie far behind the occipital condyle, so that the jaw might be as long as possible and the mouth opened widely to act as a fish trap. The stresses induced in this part of the skull by the muscles used in shutting the mouth would be met by carrying the pterygoid as far forward as possible, and by bringing its mesial border in toward the middle line so that the interpterygoid vacuity would be small. And to give the necessary flexibility the attachment of the pterygoid to the base of the brain case would be a movable articulation. The flexibility so secured would prevent any direct contact of the epipterygoid with the side of the brain case, the otic process being undeveloped and the ascending process either ill formed or at any rate not in contact with any fixed part of the skull. The condyle would not be of such a shape as to enforce a dorsoventral movement of the head to the exclusion of all others; it would in fact be round.

Since the closing in of the interpterygoid vacuity by the pterygoid bone would prevent the bulging out of the palate to allow the eyeball to be squeezed down, as in *Eryops*, the skull would necessarily be sufficiently deep, measured from dorsal to palatal surface, to accommodate the eye, and this would imply that the brain case, which actually determines this depth, would be tall.

The lack of delay in the development of bone in the primary cartilaginous skeleton would render the whole skull bony, and it would be expected to have the following structure. The basioccipital would be a large bone; therefore it might be expected to form the major part of the occipital condyle. The well-ossified basisphenoid, with its continuation backward to the basioccipital and forward to a well-developed sphenethmoid surrounding the forebrain, would have such mechanical strength that further support by an underlying parasphenoid would scarcely be necessary. Thus the basipterygoid process, to which the pterygoid was movably attached, would be completely bony and composed entirely of basisphenoid; it would of course still give attachment to the base of the epipterygoid.

The depth of the brain case, necessary to accommodate the eyes above an entirely rigid palate, would enable the front of the brain to lie high up in a cavity excavated in what would be a tall, thin sheet of bone, the brain resting on an interorbital septum. As the lower part of this septum would be thin, the parasphenoid might well be a narrow bone clasping its lower margin.

The complete ossification of the otic capsules and of the cartilaginous roof of the brain case which lay between them would provide a structure amply strong to take all the stresses arising from the action not only of the jaw muscles but of those of the neck. And this rigid brain case could be held in position by ligamentous attachment to the dermal skull roof, and by the development of ridges on the undersurface of the skull roof which would act as stops, impinging on its lateral edges. Thus the occipital flanges from the tabulars and postparietals would be absent.

As the well-ossified otic capsule would require no further support, and the occipital condyle would be formed mainly by the basioccipital, the exoccipital would be a small bone, occupying its primitive position behind the vagus nerve and not carried out into the great perichondral extensions to which it owes its great size in *Cyclotosaurus*.

These then are the qualities to be expected from the Carboniferous labyrinthodonts.

The first of them was found by Sir William Dawson and Sir Charles Lyell in the sandstone infilling of the hollow cylinder of bark of a lepidodendroid tree standing upright on a bed of coal, during their visit to South Joggins, Nova Scotia, in 1852. This animal has recently been described by my student, Miss Margaret Steen, who has shown that it has an unusual and most interesting structure. Somewhat later the anterior half of a skull was found at Pictou, Nova Scotia, and described by Sir Richard Owen, but it is probable that even earlier some fragments had been collected in Scotland.

During the 1860's T. H. Huxley described two partial skulls and two partial skeletons from the Coal Measures and the Mississippian Lower Carboniferous of Scotland and England. His accounts do not give any useful knowledge of these animals, but a series of papers published between 1869 and 1878 by a group of amateur naturalists in Newcastle-on-Tyne are of altogether different quality—clear, detailed, and in general accurate accounts of splendid material. That these papers should have been ignored for thirty years, and that during that time no one should have taken the trouble to examine the original specimens is one of the minor puzzles of paleontology.

The central figure of this group of Newcastle amateurs was Thomas Atthey. He received only an elementary education and made his living by keeping a small grocer's shop in a mining village. There he gathered fossils from the colliers, paying for them in sugar and tea so enthusiastically that he failed financially and lived for the latter part of his life on a sum of money raised by the sale of his immense collection to the museum of the local Natural History Society, which published his papers.

Atthey was a perfect preparator, removing the rock completely from all his important specimens, with a skill and care which, until the last few years, has never been equaled. The process, carried out with needles, is, as I can attest from personal experience, extremely laborious, and Atthey spent his days sitting in his shop preparing fossils. He read all the literature on the subject available in English in Newcastle, and he wrote, either alone or in association with his friends, nine papers about the structure of the fossil amphibia from a single coal seam in a single colliery. Despite some inevitable errors of interpretation these papers are admirable, correct in their statement of fact and exhibiting great morphological sense.

In 1912, on my return from South Africa, I made a first attempt to explore the evolutionary history of the labyrinthodonts in order to understand certain facts shown by the great skeletons of the capitosaurine *Uranocentrodon* which I had seen in Pretoria. This study had already led me to recognize some few of the trends which were considered in the last chapter, and the next step was to look for the ancestors of these creatures in Carboniferous times. A lucky accident took me to Newcastle and I there saw for the first time in actuality the animals which I have now to describe to you. Subsequently I have been able to describe much more material from England and Scotland, from the Lower Carboniferous as well as the Coal Measures. Professor A. S. Romer, now of Harvard University, and Miss Steen have described similar animals from the United States and from Czechoslovakia, and we have so wide a knowledge of a large and varied fauna as to be assured that we are not misled with regard to its general character by the chance discovery of an abnormal form.

The Carboniferous labyrinthodonts are in general large animals, their skulls ranging from some four to eighteen inches in length. They are varied in their skull shapes and in their adaptations and fall into two sharply separated groups, but fundamentally they are uniform in structure so far as concerns those characters which are of importance for our purpose. The structure they all possess is precisely that which the foregoing analysis has led us to expect in the ancestors of *Eryops*. In them the brain case is fully ossified, the basioccipital and basisphenoid bones which form its base being large, the exoccipitals very small, and the ear capsule massive. The whole thing is held in position with respect to the roof of the skull by lateral ridges, without the least trace of the occipital flanges which develop so greatly in the later forms. The anterior part of the brain lay within a cavity excavated in the upper part of a tall interorbital septum whose thin lower edge rested in the grooved upper surface of a narrow parasphenoid.

In the great majority of the Carboniferous labyrinthodonts the quadrate lies far behind the occipital condyle, and in all of them the pterygoid is articulated by a movable joint with a basipterygoid process which is made entirely of basisphenoid. The pterygoid extends forward on the palate to the vomers and the interpterygoid vacuity is a small thing; indeed in many forms the pterygoids of the opposite sides meet one another in the middle line (Figs. 31 and 33). In the only animal in which it is known the epipterygoid is

FIGURE 31. Drawings of a series of skulls from which the dermal bones of the lateral surface have been removed to show the pterygoid and the two ossifications, the quadrate and epipterygoid, which lie in the palatoquadrate cartilage. The series is designed to show the expansion of the ascending process of the epipterygoid and the appearance of a functional otic process in *Capitosaurus*. The forms are the osteolepid *Eusthenopteron*, the embolomerous labyrinthodont *Palaeogyrinus*, the rachitomous labyrinthodont *Eryops*, and the stereospondylous *Capitosaurus*. *Pros. Asc.*, ascending process of the epipterygoid. *Pros. Ot.*, otic process of the epipterygoid.

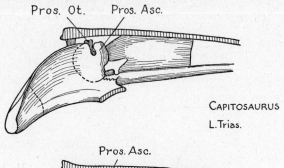

Pros. Ot. Pros. Asc.

CAPITOSAURUS
L. Trias.

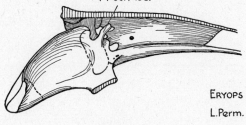

Pros. Asc.

ERYOPS
L. Perm.

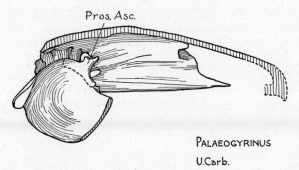

Pros. Asc.

PALAEOGYRINUS
U. Carb.

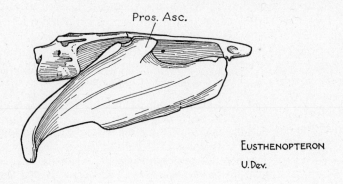

Pros. Asc.

EUSTHENOPTERON
U. Dev.

83

small, its lower end articulating with the upper surface of the basi-pterygoid process, and its ascending process being a small forwardly directed corner merely separating certain nerves from one another and having no contact with any other part of the skull (Fig. 31). Our expectations are thus fulfilled and we have evidence that our methods are sound and that the trends of evolution within the group were present in it from the beginning of Carboniferous times.

But we can go further. The British Museum contains a complete but badly preserved skeleton of an animal, *Pholidogaster,* from the Lower Carboniferous of Scotland, which is a member of our group. In Newcastle there are a skull and associated vertebrae, together with many other bones, which allow of the restoration of the skeleton of *Eogyrinus.* From the Gas Coal of Nyrany in Czechoslovakia, which is of much later age, belonging to the end of Upper Carboniferous time, I have the complete skeleton of a very small labyrinthodont of this primitive type. And from the still later Basal Permian of Texas came a very poor skeleton of *Cricotus,* the most recent member of this group, which is called from the structure of its vertebral column the Embolomeri.

It is a very curious fact, which well illustrates the extent to which paleontologists are dependent on lucky accidents for the whole of their knowledge, that these four skeletons belong to one group of the Embolomeri, the anthracosaurs, and that of the other and very much more common group, the Loxommoidea, we know nothing of any part of the body behind the head.

The skeleton of the Coal Measure *Eogyrinus* is extremely elongated. The head is deep, the section at the neck being circular; there is a short neck made inflexible because the shoulder girdle is connected to the head, as it is in fishes. The trunk is long, cylindrical, and narrows behind to pass without any sudden transition into a long, laterally flattened tail. The limbs are only incompletely known, but there is no doubt that they were short and feeble, ending in short, spreading fingers or toes. The well-developed lateral-line grooves on the skull and lower jaw show that the animal spent the greater part of its time in water, and there can be no question that it swam actively by lateral flexures of the body carried out rhythmically, so that a series of waves beginning at the head passed down to the tail, increasing in amplitude as they moved backward. *Eogyrinus* thus swam as did all the primitive fish. It must have been air breathing, because no structure exists in it which could have supported the very extensive gills necessary to enable enough oxygen to be taken in to meet the needs of so large an animal. The limbs are designed to support the body but it is evident that so long and flexible a trunk cannot have been carried free from the ground; the limbs themselves are too short and weak, and the pelvis is too feebly attached to the backbone, to have enabled the creature to move on land in any other way than by wriggling with the belly resting on the ground, as do many elongated living amphibia of similar shape though of much smaller size. In such a form of locomotion the hands and feet merely act as fixed turning points about which the animal moves. They have separate toes to give a more secure grip, and the limb bones and muscles need only be strong enough to act as struts, connecting the moving body to the fixed foot.

The track made by such an animal has the footprints of opposite sides separated by a great distance, which will be nearly, but need not be exactly, the same for the fore and hind feet. The stride, the distance between one position of a foot and the next, will certainly be short. The relative positions of the prints made by the fore and hind feet of the same side will be variable. In Fig. 32 the hind footprint is immediately in front of that of the fore foot, but it may lie on top of it or fall behind. The condition will depend upon the length of the trunk and legs, and the number of complete waves between the fore and hind legs, together with some other factors.

If the body rested on the ground the strip of sand between the footprints will be smoothed into a shallow depression which will not generally impinge upon the actual prints. Tracks of this character are actually known from Coal Measure flagstones, but they are not by any means the most common of the many types which have been described.

Nothing about the skeleton of *Eogyrinus* suggests that its ancestors had ever been animals which lived habitually on land; it is in all probability a primarily aquatic form directly derived from fish, traveling over land only for short distances to pass from one pond to another. But the other embolomerous amphibia had other habits. The small animal *Diplovertebron*, from the Gas Coal of Nyrany, shows itself, by the absence of lateral-line grooves in its skull and by the long, slender digits and long, powerful legs, to have been terrestrial. The development of the necessarily powerful musculature of the shoulder has led to a change in the character of the clavicle or collarbone. In fish this is a flat sheet of bone in the skin covering nearly the whole of the so-called primary shoulder girdle, with which the arm, in them a pectoral fin, articulates and from which its musculature arises. This

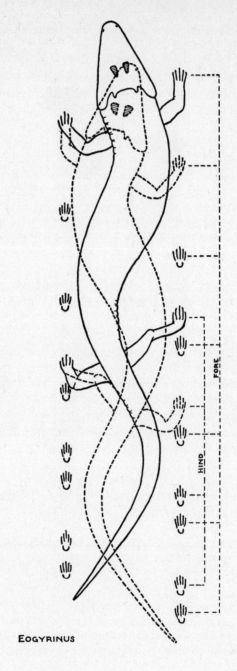

EOGYRINUS

condition was carried on by direct inheritance to *Eogyrinus*. In *Diplovertebron,* however, in order to expose a larger surface of the primary shoulder girdle for the attachment of muscles, the upper part of the clavicle shrinks down until it becomes a slender rod, the smallest element capable of carrying out the mechanical purpose which the bone serves. In this way the type of collarbone found in *Eryops* and all other terrestrial labyrinthodonts came into existence. When these gave rise to the secondarily aquatic *Cyclotosaurus* and similar animals of the same period, the lower end of the clavicle was greatly enlarged for some mechanical reason connected with the width of the very flat body.

It is thus highly interesting to find, even in rocks of Lower Carboniferous age, clavicles of this secondarily aquatic type, which are to be distinguished from those of Triassic forms only because they are fitted to a body of cylindrical shape.

That the Embolomeri are actual ancestors of the later labyrinthodonts has been made certain by Miss Steen's description of the Upper Coal Measure *Dendrerpeton.* This animal has the single condyle of an embolomer; its brain case is well ossified and the occipital lappets of the tabulars and postparietals are rudimentary, not nearly reaching the small exoccipital. The basipterygoid process is made by the basisphenoid and has a movable articulation with the pterygoid, but the epipterygoid is similar to that of *Eryops.* The quadrate lies far back and the interpterygoid vacuity is present but small, the pterygoid extending to the vomer. In these details and in several others the animal is a perfect intermediate between the Embolomeri and the rachitomous contemporaries of *Eryops* (see Fig. 33).

Thus it is certain that labyrinthodonts must have had a history extending far enough back before Carboniferous times to have allowed a primitively aquatic animal to produce terrestrial descendants, and for these in turn to give rise to animals which returned to an aquatic life.

The existence of *Eogyrinus* shows that the first amphibians, though air breathing, were aquatic, and that they soon became possessed

FIGURE 32. A restoration of *Eogyrinus,* a Coal Measure labyrinthodont, walking, to show that its method of locomotion involves body movements identical with those used in swimming by all primitive fish, including the osteolepids and Dipnoi which are close relatives of the fish ancestors of the amphibia. The figure in broken line is the position at the beginning of a stride, that in continuous line at the end of the same movement. The character of the tracks is shown.

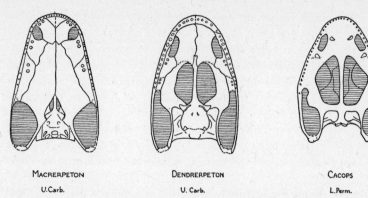

MACRERPETON   DENDRERPETON   CACOPS

U.Carb.   U. Carb.   L.Perm.

FIGURE 33. This drawing is intended to show that intermediate forms are known between typical Carboniferous Embolomeri and typical Permian Rachitomi. The three animals, *Macrerpeton* from the Coal Measures of Linton, Ohio, *Dendrerpeton* from the Coal Measures of Nova Scotia, and *Cacops* from the Lower Permian of Texas, were selected because they have skulls of much the same contour, though they are not members of the same phylogenetic line. Note that in the size of the interpterygoid vacuities and in the mode of articulation of the pterygoid with the brain case *Dendrerpeton* is intermediate between the two animals between which it is placed. *Macrerpeton* after Romer; *Dendrerpeton* after Steen; *Cacops* after Williston.

of that hereditary mechanism which secured their evolution through the carrying out of trends which persisted unchanged while the stock changed from water to land and back to water again. The only escape from this conclusion would be to assume that these trends only arose in some embolomer which had become terrestrial. It is therefore necessary to exclude this possibility. The only way in which this can be done is by investigating the other orders of amphibia, the Adelospondyli, Lepospondyli, and Phyllospondyli, and the recent amphibia.

It is convenient to begin with the Phyllospondyli which, until recently, were known to us only by a group of small animals, each six or seven inches long, whose remains are found commonly in rocks at the extreme top of the Carboniferous and of Lower Permian age in Germany, France, Czechoslovakia, and Hungary. The typical branchiosaurs have recently been redescribed by my students, Drs. Bulman and Whittard. In 1909 the late Dr. R. L. Moodie described, though only incompletely, some branchiosaurs from the Coal Measures of the United States; and I had a much more satisfactory specimen of somewhat earlier period from England. But a few years ago

Professor Romer and Miss Steen, independently and simultaneously, gave admirable and well-illustrated accounts of a famous Pennsylvanian fauna from Linton, Ohio, which for the first time made us aware that the Phyllospondyli of that time were a varied and extensive group.

If we compare a Linton animal, *Pelion,* which is the most advanced of its group, with *Branchiosaurus,* we find that in this group, as in the labyrinthodonts, the interpterygoid vacuities widen and the quadrate moves forward; that indeed the evolutionary changes, so far as they can be seen, follow definite trends which are common to both groups. A bulk comparison of all the Linton Phyllospondyli with all those of Permian age shows that in them both the flattening of the head and delay in onset of ossification of the cartilaginous skeleton occur exactly as in labyrinthodonts.

It is therefore necessary to discover the origin of this group. Recent still unpublished work of Miss Steen has shown that certain animals, which are known to be related to the Phyllospondyli, as interpreted above, have, when adult, a vertebral column which can only be taken as a derivative of the embolomerous type. The skull pattern and indeed all other skeletal characters of this group are also capable of derivation from those of the early Carboniferous labyrinthodonts, and there is very little doubt that this is actually their origin. If so, indeed in any case, the branchiosaurs present a most interesting example of an accelerated evolution; by Lower Permian times they had acquired a structure which the more conservative labyrinthodont stock achieved only by the Trias.

The Lepospondyli are small animals which are still rather little known, most of them remarkable for their possession of bizarre specializations. One family is composed of legless, wormlike animals. Another contains greatly elongated creatures with extremely small limbs, an elongated, sharp-pointed skull, and a laterally flattened swimming tail. A third has a horned skull, a cylindrical body of moderate length, and a long tail which seems to have been derived from a swimming appendage that has undergone reduction and become nearly circular in section.

Only in the last family do we know enough about a series of genera to trace the evolutionary changes which took place during the history of this group. The earliest forms, *Ceraterpeton,* from Ireland and England, and the similar *Diceratosaurus,* from Linton, Ohio, are small animals with a deep, wedge-shaped skull, which already possessed two separate exoccipital condyles but had small interpterygoid vacuities and a movable articulation between the pterygoid and the basipterygoid process. The posterolateral corners of the roof of the skull are pulled backward into a sharp-pointed, powerful spine. *Batrachiderpeton,* an English contemporary of *Ceraterpeton,* has a much more developed horn and leads on toward the American Lower Permian *Diplocaulus.* This animal (Fig. 34) has an immense

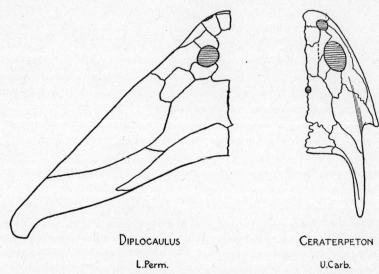

DIPLOCAULUS   CERATERPETON

L.Perm.   U.Carb.

FIGURE 34. The dorsal aspects of the skulls of two lepospondyls. To show the very characteristic pattern associated with the long posterior horns. Only the right side of the skull of *Ceraterpeton* from the Coal Measures of Ireland and the left of *Diplocaulus* from the Permian of Texas are shown. *Ceraterpeton* after Steen, *Diplocaulus* from Williston.

head carried out into huge lateral horns, and its occiput is perhaps the most depressed known in any amphibian. The body is narrow compared with the head and must have been very much flattened in front, but farther back it becomes round and finally tapers away into a slender, almost whiplike tail. The legs are not disproportionate to the body although they are dwarfed by the immense head. *Diplocaulus* (Fig. 35) has large interpterygoid vacuities, and the pterygoid is attached by a strong interdigitated suture to the widened hinder end of the parasphenoid and to the exoccipitals, in exactly the manner characteristic of the Upper Triassic labyrinthodonts. The basioccipital, basisphenoid, otic bones, and supraoccipital bone have vanished entirely, or at least become extremely reduced, and the exoccipital has spread outward as a perichondral bone. Thus in this group also we have a course of evolution exactly like that which occurs in all labyrinthodonts and in the Phyllospondyli. The relationship between the Lepospondyli and the Embolomeri cannot be close; at the first (nearly simultaneous) appearance of the two groups in the Scottish Lower Carboniferous they differ as greatly as they do at any subsequent time.

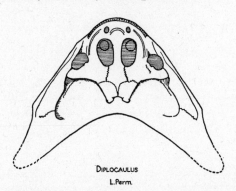 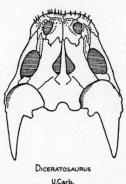

DIPLOCAULUS
L. Perm.

DICERATOSAURUS
U. Carb.

FIGURE 35. The palates of two lepospondyls. To show the very great expansion of the interpterygoid vacuities and the attainment in *Diplocaulus* of a mode of attachment of the pterygoid to the base of the brain case identical with that which appears in the latest labyrinthodonts such as *Cyclotosaurus* and *Plagiosaurus*. *Diplocaulus* from the Texan Permian, after Case; *Diceratosaurus*, which is nearly identical with *Ceraterpeton*, after Jaekel's figure of a specimen from the Coal Measures of Linton, Ohio.

The only remaining group of Carboniferous amphibia, the Adelospondyli, is still so incompletely known that it is impossible to say how far its evolutionary history conforms to that of the other groups. The two Lower Carboniferous forms known differ greatly from the Embolomeri.

There is thus evidence that in Lower Carboniferous times, at what was till very recently the period of their first occurrence, the amphibia were a highly varied group comprising three independent orders and exhibiting great variety of adaptation. It is therefore certain that they must have had a long history in Devonian times, and the fact that no trace of them has ever been found in the widespread and well-searched continental deposits of the Upper Old Red Sandstone was and remains a mystery.

The recent discovery by a Danish government expedition under Dr. Koch of skulls of amphibia in the Upper Devonian of Greenland was thus in no way unexpected, and the publication of a "preliminary" description of them by Dr. Säve-Söderbergh of Stockholm raised my hopes that at last we should see the ancestors of the Embolomeri and confirm or refute the views which I had expressed on

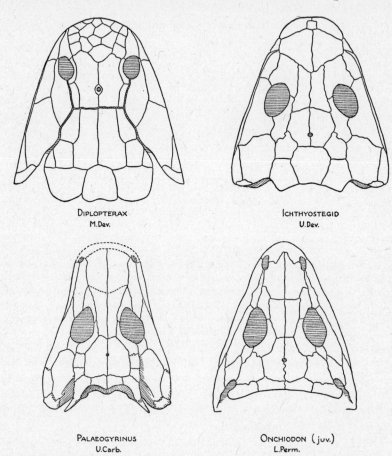

DIPLOPTERAX
M.Dev.

ICHTHYOSTEGID
U.Dev.

PALAEOGYRINUS
U.Carb.

ONCHIODON (juv.)
L.Perm.

FIGURE 36. The dorsal surface of the heads of one fish and three amphibians, to show that it is possible to institute a comparison between the patterns made by their dermal bones. It will be seen that the Upper Devonian ichthyostegid is in some ways less like the fish than is the later *Palaeogyrinus*.

Diplopterax is an ostelepid fish from the Middle Old Red Sandstone (Devonian) of Scotland; the figure is an original reconstruction by Dr. T. S. Westoll. The ichthyostegid is a synthetic drawing from those published by Säve-Söderbergh. *Palaeogyrinus* is from Watson, 1926; and *Onchiodon* is from a very juvenile individual, from the Rothliegende of Neiderhässlich near Dresden, in the author's collection.

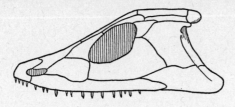

ONCHIODON (juv.)
L. Perm.

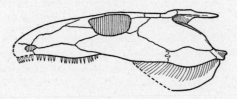

PALAEOGYRINUS
U. Carb.

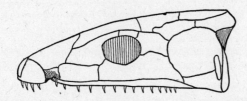

ICHTHYOSTEGID
U. Dev.

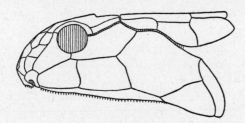

DIPLOPTERAX
M. Dev.

FIGURE 37. Side views of the skulls shown in Fig. 36. To show especially the identical position of the external nostril on the margin of the mouth in *Diplopterax* and the ichthyostegid.

93

the relationships of these animals, and of the amphibia generally, to the osteolepid fish. In many ways my expectations were confirmed, but the knowledge we now have of these animals, which Dr. Säve-Söderbergh called ichthyostegids, raises as many new problems as it solves.

Comparison of Figs. 36 and 37, which place side by side skulls of a synthetical ichthyostegid and an osteolepid fish, shows at once the possibility of a not very remote relationship between the two groups. Although the pattern of the dermal bones on the roof of the head of the two animals is actually different, it is nonetheless reducible to the same basal plan. In each a chain of three bones crosses the posterior part of the skull. In each a chain of bones passes forward from the lateral element of this series over the eye to bow outward toward the nostril and then turn inward as the upper part of the premaxilla. In each animal two areas—one behind the other in front of the eye—which lie between these chains of bones are filled in by others, differing in number but varying so greatly in their arrangement in different individual osteolepid fish as to show that the variation is not of very great significance. In each animal the eye is surrounded by a circle of five bones, one of which is separated from the nostril by a single element, the septomaxilla. This series of circumorbitals rests on the maxilla ventrally and is continued backward over the cheek by three bones: the squamosal, the quadratojugal, and the preoperculum.

The palate of the amphibia is built on the same pattern as that of the fish (Fig. 38); the same membrane bones are present in each and they have the same relations to one another. Indeed, the only important difference is that in osteolepid fish the basipterygoid process articulates solely with the epipterygoid, while in the amphibian the pterygoid grows inward to gain a direct articulation with it. One interesting point of resemblance lies in the structure of the nostrils. Both fish and amphibian have an external nostril, a notch in the border of the mouth separated from the internal palatal nostril only by the inturned anterior end of the maxilla, which in the fish at any rate underlies a splint of bone belonging to the fully ossified nasal capsule. In later osteolepids the internal nostril retains its original form as a triangular hole, bounded by the premaxilla and maxilla laterally and the vomer and palatine mesially, but the external nostril migrates until it lies high up on the side of the snout. An identical change takes place in the amphibia, where the external nostril, although it never moves so far away from the mouth as in

the fish, is in all later forms completely cut off from the border of
the mouth. In the Coal Measure embolomeran *Orthosaurus* the external nostril is connected to the internal nostril by a smooth but shallow groove, the last relic of their former close connection, indeed
of their formation by the secondary division of an originally single
pit.

The ichthyostegids thus provide the expected evidence of relationship between the amphibia and the osteolepid fish. It seems
evident that the two groups had a common ancestor more nearly
resembling these fish than it did the Dipnoi, the only alternative.

Although they retain primitive characters in many parts of their
structure it is certain that the ichthyostegids are in other ways too
advanced to be in any way ancestral to the embolomerous labyrinthodonts. Before their systematic position can be discussed in-
telligently it is necessary to determine how great is the difference
in age between the two groups.

The work of the Danes has shown that the Upper Devonian amphibia of Greenland come from the extreme summit of that system.
They lie in the upper of two life zones which succeed the uppermost
horizon of the Old Red Sandstone found in Scotland and are thus
considerably younger than any horizon there which has yielded an
appreciable number of determinable fish. In the south of Scotland
the Upper Old Red Sandstone passes upward conformably into the
base of the Carboniferous, there being so complete and perfect a
transition from one to the other that any dividing line is entirely
arbitrary; indeed it is drawn locally at the highest bed from which
Holoptychius scales have been collected. The Burdiehouse limestone,
the lowest horizon from which amphibia have been obtained in
Scotland, lies low in the Lower Carboniferous, not far above its
arbitrary base. It is thus probable that the Greenland Upper Devonian amphibia are only very little older than the Lower Carboniferous forms from Scotland; there is indeed no assurance that they are
any older at all.

If we compare (Figs. 36, 37, and 38) the skull of an ichthyostegid
with that of an embolomer, directing our attention to those qualities
whose evolution we have traced up to the Upper Trias, we find
that the skulls of both are deep but the quadrate lies much farther
forward in the ichthyostegid than it should. The interpterygoid
vacuities are very small and the pterygoids articulate movably with
the brain case in both. The occipital condyle is single. But so far as

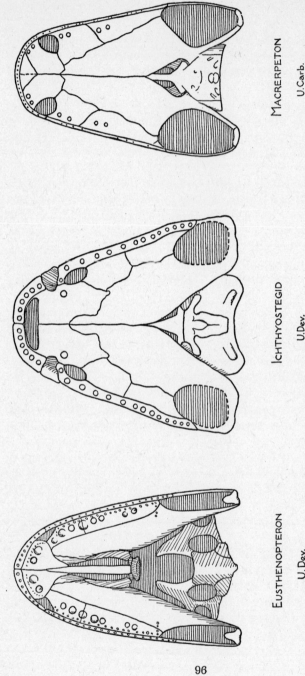

EUSTHENOPTERON
U.Dev.

ICHTHYOSTEGID
U.Dev.

MACRERPETON
U.Carb.

FIGURE 38. The palates of the Upper Devonian osteolepid fish *Eusthenopteron*, the Upper Devonian amphibian ichthyostegid, and the Upper Carboniferous embolomeran *Macrerpeton*. To show the close similarity in detail as well as in general plan between these three groups in the region. The ichthyostegid is more primitive than the others in retaining an external nostril which notches the border of the upper jaw. In other characters, the anterior situation of the quadrate condyle and the presence of descending flanges from the tabulars, it is very advanced. *Eusthenopteron* after Watson; ichthyostegid an interpretation of Säve-Söderbergh's figures; *Macrerpeton* after Romer.

96

can be judged from the published photographs, the otic capsule of the ichthyostegid was unossified or at most contained very little bone; and no doubt following on this lack of ossification a special process, homologous with the occipital lappet of *Eryops,* passes down to the otic capsule from the lower surface of the tabular. Thus in some very significant features the ichthyostegids are far more advanced than the Embolomeri.

The dermal skull pattern of the ichthyostegids differs so widely from that of the labyrinthodonts that there can be no close relationship between the two groups. There is no reason to suspect any close relationship to the Phyllospondyli, Lepospondyli, or the Adelospondyli. Thus the ichthyostegids appear to be a short-lived side branch, from the base of the amphibian stock, which retained some very primitive characters but combined them with structures formed by a greatly accelerated evolution following in part the trends familiar in other lines of amphibia.

There remain for consideration only the groups of amphibia which have still living representatives. The most vigorous of these orders, that which contains the frogs, is an example of a group exhibiting an extremely active adaptive radiation following on a period of obscurity during which, by the end of the Jurassic time, a very characteristic and altogether peculiar body form and structure was achieved. Within the past few months Dr. J. Piveteau of Paris has found in those Basal Triassic rocks in Madagascar long famous for fossil fish a single specimen which is with complete certainty an ancestral frog.* It is not necessary to discuss it for my present purpose. The small group of the Coecilia has no known fossil representative. The few fossils known of urodele amphibia are of little importance; they merely show that modern genera date back into Tertiary times, and that the earliest known fragments of late Jurassic age do not differ significantly from those which still exist. But these forms are so varied in character and have been so fully studied that they are of great service in all attempts to interpret the evolution of fossil amphibia.

The older naturalists divided the living tailed amphibia into two groups. One, the Perennibranchiata, which they naturally believed to be the more primitive, contains only exclusively aquatic animals, in which a series of pairs of pinnate external gills rose from the neck. In the other, more advanced in structure, the external gills found in

* See Bibliography, 78.

the larva of all living amphibia shriveled up during the course of the metamorphosis by which the aquatic larva became transformed into an adult which might be terrestrial or might remain all its life in water. Included in the former group was the axolotl (Fig. 39), which grows to a length of some eight or nine inches and breeds freely, laying eggs from which hatch larvae which by mere growth, without metamorphosis, become adult. About fifty years ago it was found that some, though not all, axolotls could, by mere shortage of oxygen in the water in which they lived and by being compelled to come onto land, be caused to metamorphose into terrestrial salamanders.

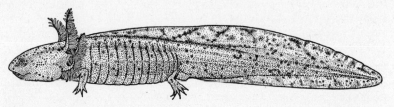

FIGURE 39. The axolotl, to show the character of the external gills in the persistent larva of a living urodele.

Much more recently it has been shown that all axolotls can be caused to metamorphose immediately by a single injection, or meal, of thyroid gland of a dog or any mammal or indeed of any vertebrate, including the axolotl itself, if sufficient gland be given. The failure of the axolotl to metamorphose is in fact due to the incapacity of its own thyroid to produce enough thyroxin to enable the concentration of that substance in the blood to be raised above the threshold. The adult, sexually mature axolotl differs in no important respect from any true perennibranchiate urodele, and attempts were therefore immediately made to transform these animals by administration of thyroid. All were unsuccessful. These animals have lost the power possessed by other amphibia of responding to thyroid by metamorphosis.

The obvious interpretation of the Perennibranchiata is that they are permanent larvae, that they represent the result of a postponement of the metamorphosis to a later and later period in life. That this view is correct is confirmed by every recent study of their comparative anatomy. The most striking result of the delay is the retention of the external gills. These coexist with lungs which are functional as respiratory organs, and appear early in the development of the larva. External gills are quite different from the internal gills found in all fishes, and no structures intermediate between the two occur. They are found in the larvae of all living amphibia, in those of two of the three living genera of dipnoan fishes, and in the larvae of the curious, closely allied African fish, *Polypterus* and *Calamoichthys*. In all these cases the external gills are associated with a large yolky egg which (in all but the first) is laid in large masses in shallow tropical water; in the only case where analyses have been made the water proved to be extremely poor in oxygen. The gills are in fact larval organs introduced to carry the small fish over the difficult period when their oxygen requirements have become considerable, the lungs are not yet functioning, and the internal gills are either undeveloped or inadequate.

It is fortunate that we have direct evidence that labyrinthodonts and branchiosaurs had a metamorphosis. Drs. Bulman and Whittard have recently confirmed and extended in *Branchiosaurus* the account of this process given more than forty years ago by H. Credner. In material contained in extremely fine-grained, fresh-water limestones of Lower Permian age from Odernheim in the Palatinate, branchiosaurs of all sizes occur in perfect preservation, the soft parts being shown as acid-resisting carbonaceous films. In individuals with a total length of seven or eight centimeters (Fig. 40) these animals always show four series of rows of minute denticles which form loops in the side of the neck, extending upward from a point near the ventral line to the level of the vertebral column. The upper end of each loop turns a little inward. The front loop is a single row of backward-pointing denticles. The next two are each double, the points of the denticles being directed forward and backward. In the fourth loop there is a single series of forward-directed denticles. Comparison with recent urodele larvae and with fish shows beyond all question that these structures are gill rakers and that together they guarded the inner openings of three pairs of gill slits. The depth of these gill slits is so great that it is highly probable each contained an actual internal gill, a direct inheritance from fish ancestors. But in addition Bulman and Whittard have found in many specimens (see Fig. 40) three pairs of external gills—long, slender processes rising from the upper part of the body above and in the region of the gill slits. They have shown that during a very long period of metamorphosis the gill rakers first disappear, the external gills remaining, but eventually the latter too are absorbed, not being retained by the full-grown adult.

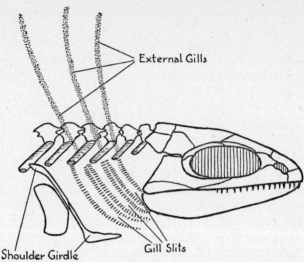

FIGURE 40. Reconstruction of the anterior end of the skeleton of the phyllospondyl *Branchiosaurus*. The drawing represents a larva, with a skull length of about 1 cm. It shows the series of minute pointed denticles, the gill rakers, which were attached to the inner surface of the visceral arches and guarded the very long gill slits, of which there were three pairs.

The external gills are shown as they appear in the specimens from Odernheim, Pfalz. They seem to be attached each to a visceral arch in front of an open gill slit. During the prolonged metamorphosis the gill rakers disappear first, the external gills being resorbed only after an interval. Original from the material used by Bulman and Whittard.

In 1856 von Meyer showed that the Lower Permian labyrinthodont *Archegosaurus,* when small, possessed loops of gill rakers extremely like those of *Branchiosaurus,* and that these vanished when the skull reached a certain length. The preservation of *Archegosaurus* is such that no trace of external gills could be expected to be preserved.

Only one larva of a later labyrinthodont is known, a specimen from the Lower Triassic Hawkesbury Sandstone of Gosford, New South Wales, described by Stephens as *Platyceps wilkinsoni.* This is a brachyopid, probably near to *Batrachosuchus.* The little animal, with a skull about three centimeters in length, possesses a series of well-ossified branchial arches tightly packed between the skull and the shoulder girdle. The upper ends of these arches are far apart, and from the summit of each a relatively long, ill-preserved but apparently bony axis projects. These can only be the supports of external gills. This specimen is clearly young, a larva which might have lost its gills later; but fortunately a skeleton of *Dwinasaurus* so large (about four feet in length) that it must be adult retains a complete series of massive branchial arches, only explicable if they supported external gills (Figs. 41 and 42). We have thus excellent evidence that the labyrinthodonts, like the urodeles, ultimately developed into perennibranchiates; that, in fact, in this respect, as in others, the history of the two groups was the same.

This circumstance is of the utmost importance because it gives us the only opportunity I know of studying in living material and by experimental methods the mechanism involved in the development of an evolutionary trend common to all members of a class of the animal kingdom.

APPENDIX

The description of the structural characters that would be expected in an ancestor of *Eryops*, which occupies the first two or three pages of this chapter, was in 1937 a prediction of an unknown animal. In 1942 Romer and Whitter described as *Edops* an amphibian occurring in deposits in Texas which lie below those in which *Eryops* is an exceedingly common animal. In these deposits *Eryops* itself is rare, and it is fully evident that *Edops*, though not its ancestor, retains much of the structure it possessed. In every point *Edops* agrees precisely with the predictions made in this chapter, a satisfactory confirmation of the validity of the arguments on which it was based.

As I have stated, the large Carboniferous labyrinthodonts fall into two sharply separated groups, which have recently been shown to be in effect ancestral to different groups of labyrinthodonts and their allies. The group of the loxommids, apart from certain obvious peculiarities, seems to have been close to the ancestry of the true labyrinthodonts, while the other of the two groups, that of the Anthracosaurs of which *Eogyrinus* is a member, is related to the seymouriamorphs as well as to the Texan Permian *Archeria* (*Cricotus*). But in Carboniferous times the two groups were fundamentally similar in structure. Unfortunately the postcranial skeleton is still known only in Anthracosaurs, no bone, not even a vertebra, being definitely associated with any skull of the much commoner loxommids.

The animals which in 1937, following Romer and Steen, I referred to the Phyllospondyli are clearly, in many cases, of different sys-

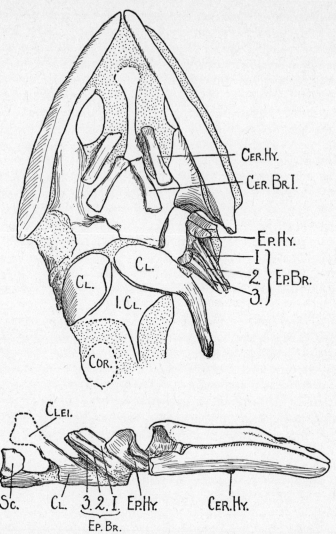

FIGURE 41 and FIGURE 42. Ventral and lateral views of the anterior part of the skeleton of the Upper Permian brachyopid labyrinthodont *Dwinasaurus* from the Dwina River, Russia. To show the persistence into an adult of a powerful, well-ossified series of branchial arches, only explicable as the supports of large external gills. The specimen gives evidence that in labyrinthodonts, as in urodeles, the process of evolution involved a delay and final suppression of metamorphosis. From a cast in the British Museum of a specimen now in Moscow. *Cer. Br.* 1, ceratobranchial; *Cer. Hy.*, ceratohyal; *Cl.*, clavicle; *Clei.*, cleithrum; *Cor.*, coracoid; *Ep. Br.* 1,2,3, epibranchials; *Ep. Hy.*, epihyal; *I. Cl.*, interclavicle; *Sc.*, scapula.

tematic position. It has been suggested by Romer that most of them do not belong to the labyrinthodonts at all, but may be brigaded with the so-called Upper Devonian ichthyostegids, described by Säve-Söderbergh. This view is plausible but it still rests, I feel, on slender foundations. In any case these creatures are very unlike normal labyrinthodonts in many features; their vertebral column, for instance, although its structure is not yet satisfactorily known, seems to include paired paranotochordal elements not conforming to those found in normal labyrinthodonts. Nothing of importance has recently appeared on Lepospondyli or Adelospondyli.

The ichthyostegids from Greenland still await further description, but Westoll has described as *Elphistostege* part of the upper surface of a skull which comes from the well-known fish-bearing deposits of Scaumenac Bay. These beds lie somewhere about the passage from the Middle to the Upper Devonian rocks of Greenland which yield ichthyostegids. He has shown that the skull fragment provides, fundamentally with regard to the relative sizes of the pre- and postorbital region of the head, an intermediate between the osteolepid fishes in which the occipital region is very long and ichthyostegids in which it is relatively shorter though even then much longer than it is in normal labyrinthodonts. The matter has a great technical importance because it has led Westoll to an entirely new (and obviously correct) interpretation of the relation between the dermal bones of the head in osteolepids and in tetrapods, and has gone far to confirm the descent of the tetrapods as a whole from a fish ancestor, which was either an osteolepid or some closely related form.

The discussion of ichthyostegids in Chapter IV was, of course, based on our knowledge at that time. If we may accept Romer's new view that *Colosteus* and *Erpetosuchus* are closely related to Ichthyostega, then we must recognize that this group had an evolutionary history which, so far as it continued, agreed with that of the labyrinthodonts. For the palate of *Erpetosuchus,* as it has been described by Steen and Romer, has large interpterygoid vacuities, the pterygoids not reaching the vomers and indeed only just touching the palatines. The quadrate condyle lies essentially in the same transverse plane as the occipital condyle, though the pterygoid still articulates movably with the basipterygoid process.

Since 1937 Dr. J. Piveteau has published an additional description of the Lower Triassic frog from Madagascar, which he calls *Protobatrachus,* and I have given a further account of the structure of its skull. Furthermore I have described the structure of a small amphibian, *Miobatrachus,* from the Mazon Creek beds, of high Coal Measure (Pennsylvanian) age. This animal shows in its structure much resemblance to *Protobatrachus,* and in fact forms a good morphological ancestor to the whole group of frogs.

It is itself capable of derivation from the much older Lower Coal Measure *Eugyrinus,* a creature which is quite clearly of labyrinthodont derivation but which, in some of its characters, seems to be definitely related to the true branchiosaurs. The interpretation of *Dwinasaurus* as a perennibranchiate labyrinthodont, which I first put forward in 1926, has been very completely confirmed, and the evidence on which it rests was elaborated by Bystrow in 1938 and 1939.

V. EARLY REPTILE HISTORY

IN THE LAST three chapters I have discussed the evolutionary history of an entire class of vertebrates, the amphibia. The group is remarkable in that throughout, in each order in which sufficient facts are known, it can be shown that the more fundamental changes take place steadily with time, uninfluenced by the habits of the creatures themselves. Superimposed on these basal changes are, of course, others which fit special groups of animals for specific types of life. The structures which so change in accordance with definite trends are numerous, but it is possible by consideration of the mechanics of the animal to show that many of them may be interpreted as the necessary results of other changes which have to be made in order that the skull may remain structurally sound throughout the history of a stock.

Finally we were left with but two trends, whose following out inevitably involves the other changes in structure, and these trends are common to most amphibia. Thus the history of the group presents innumerable cases of parallel evolution on the largest scale.

This phenomenon is very widespread; the work now in progress on the Devonian, Carboniferous, Permian, and Triassic fish belonging to the group Actinopterygii by Professor Stensiö and his associates, by Dr. J. Piveteau in Paris, and by Dr. Westoll and Dr. Brough who work with me, is showing that the whole history of that group involves the often-repeated occurrence, at many different periods, of changes in structures of a definite kind carried to a greater or less extent.

The fact that the whole history of the amphibia is essentially the working out of a few definite evolutionary trends suggests that the next step in any consideration of the mechanism of their evolution should be an investigation of the mode of origin of the reptiles, which are undoubtedly the descendants of some amphibian. The most important distinction between the two groups is their life history.

All amphibia, both living and extinct, appear to have laid their eggs in water. Even thoroughly terrestrial living forms do so, sometimes returning to water for that purpose only. The few living amphibia (the Coecilia, certain salamanders, and frogs) that lay their eggs on land form no real exception to this statement, because in them the special devices introduced to enable development to proceed are of such a nature that it is at once evident they in effect provide a pool of water in which the larva can (at any rate in principle) swim.

The amphibian egg is large and extremely similar in all its characters to those of the more primitive living fish. It is enclosed within a gelatinous capsule and development takes place therein until, from materials contained in the egg, there is formed a larva capable of swimming and feeding; this then hatches by digesting a hole in the capsule. The larva lives in the water, breathing by external gills, abstracting oxygen from the water, and feeding and growing actively. Finally it changes by a sudden metamorphosis into the adult condition, often altering its food and other habits completely. This mode of development makes it necessary for amphibia to return to water during the breeding season and renders it difficult for them to invade and colonize deserts.

The reptiles escape this necessity. In them the egg is fertilized internally; it is then, during its passage down the oviduct, coated with a very dilute watery solution of a protein like the white of a hen's egg, and surrounded by a shell which is always tough and may be calcareous. Such an egg can carry out the whole of its development on land if it is kept at a sufficiently high temperature, and if the surrounding atmosphere is moist enough so that evaporation from the surface of the shell will not dry the whole thing up. The "white of the egg" performs two functions: it protects the living yolk, the actual egg, from damage of all kinds; and it provides a store of water sufficient to make the necessary dilution of the concentrated food substances contained in this yolk. The actual food value of the white is negligible, but it is possible that it may also be a reservoir of mineral salts. The creature which hatches from a reptile egg is capable of living a life similar to that of the adult; it feeds on the same sort of food and grows without any metamorphosis into the adult. It should be pointed out that a reptile egg must be laid on dry land; aquatic reptiles—turtles, tortoises, crocodiles, lizards, and snakes—lay their eggs on land, if they have not become viviparous.

The few existing types of reptiles have a skeleton exceedingly unlike that of the living amphibia, and it is only when we trace the groups back to Paleozoic times that they present any similarity. It is therefore, to an evolutionist, most pleasing that there is a group of animals whose structures are very completely known, which are of uncertain position and may be reptiles or may be amphibia.

The best-known genus, *Seymouria,* was originally found and described by Professor Broili of Munich, who held it to be a reptile. It comes from the Basal Permian of Texas, where it is rather rare. Seven years later Professor Williston of Chicago obtained and described a complete skeleton contained, unfortunately, in a very hard nodule; he also considered it to be a reptile. Four years afterward I collected an imperfect but very well-preserved skeleton and also described it as reptilian. Then in 1922 Dr. Broom of South Africa discussed *Seymouria* and concluded it was an advanced embolomerous amphibian, and in 1925 Professor Sushkin of Leningrad brought forward new evidence, drawn largely from an allied Russian animal, which led him to support Dr. Broom's interpretation. In 1928 Dr. Romer, then of Chicago, summed up the evidence and came down on the side of reptilian affinity.

It seems evident that any animal which can attract such international interest and lead equally competent students to such different conclusions should be capable of throwing light on the problem of the process whereby a new class came into existence.

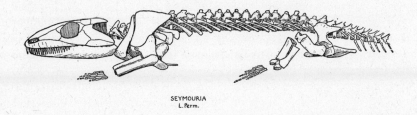

SEYMOURIA
L. Perm.

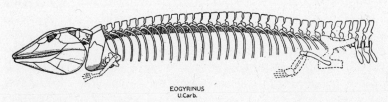

EOGYRINUS
U. Carb.

FIGURE 43. Reconstructions of the skeletons of two animals, *Eogyrinus,* a Coal Measure embolomerous labyrinthodont; and *Seymouria,* "the most primitive known reptile," from the Lower Permian of Texas. To show the powerful limbs of *Seymouria* associated with its terrestrial habits, and the short vertebral column with its extremely massive neural arches. Both after Watson.

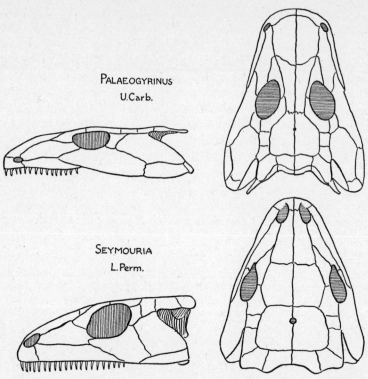

PALAEOGYRINUS
U.Carb.

SEYMOURIA
L.Perm.

FIGURE 44. Dorsal and lateral aspects of the skulls of *Seymouria* and of *Palaeogyrinus*, a Coal Measure embolomerous labyrinthodont. To show the identity of skull pattern and the minor differences of shape. After Watson.

Seymouria is a small animal about two feet long, looking like a very heavily built lizard, with short, thick legs and no neck (Fig. 43). The upper and lateral surfaces of the skull are identical in pattern with those of certain embolomerous amphibia (Fig. 44). Like these creatures *Seymouria* has on the side of its head behind the eye a deep bay, the otic notch, across which the tympanic membrane of the ear was stretched. In both groups the end of the stapes was attached to this membrane in order that the vibrations into which it was thrown by sound waves falling on it might be transmitted to the inner ear, there to activate the actual sensory receptor cells. In one surface feature *Seymouria* is different from the majority of Embolomeri; the bone called the lachrymal extends from the orbit to the nostril, retaining a primitive condition found in fish and in ichthyostegid amphibia but lost by most others.

The palatal surface of *Seymouria* (Fig. 45) is equally like that of an embolomeran; it is composed of the same bones, arranged in the same way, and bearing in the same places teeth having the same microscopical structure. In the brain case and occiput, however, there are significant differences. The occipital condyle of *Seymouria*, though like that of an embolomeran it is single, nearly circular, and composed of basioccipital and exoccipitals, is convex; it is a rounded knob in contrast to the concave, fishlike amphibian structure. In both the exoccipital is a small bone, but in *Seymouria*, as in *Dendrerpeton*, the Coal Measure amphibian which is intermediate between

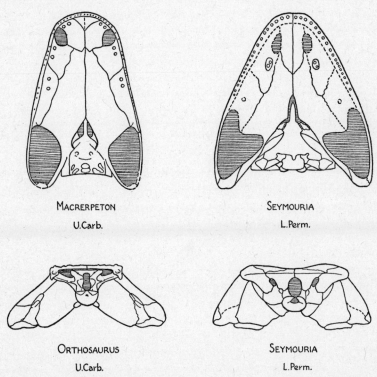

FIGURE 45. The palate and occiput of *Seymouria* compared with those of two Embolomeri, from the Coal Measures. To show the great structural resemblance and the differences between the two animals. In *Seymouria* there are flanges, descending into the occiput from the membrane bones of the skull roof, which are absent in the Embolomeri. The occipital condyle is rounded instead of being concave, and there are powerful ridges ending in the basisphenoidal tubera not shown in the amphibians. *Macrerpeton* from Romer, the remainder after Watson.

the embolomeri and later labyrinthodonts, there are small occipital flanges projecting downward as lappets from the tabulars and post-parietals. The way in which the basisphenoid is drawn out into basipterygoid processes, which articulate with the pterygoid, is the same in both groups. But in *Seymouria,* as Sushkin showed, the fenestra ovalis is carried out laterally to a place below the tabular very near to the tympanic membrane. This arrangement is unique; nothing like it occurs in any other animal, amphibian or reptilian. But this strange structure occurs in association with a build of the ventral surface of the basisphenoid unlike that in any Carboniferous amphibian, though capable of derivation from such, and the structure is identical with that found in many reptiles, in *Captorhinus* and *Varanosaurus* for example. Thus so far as the skull is concerned there is little reason for regarding *Seymouria* as other than a some-what peculiar embolomerous labyrinthodont.

In the lower jaw the two are equally similar; *Seymouria* differs slightly in that a bone called the splenial is entirely on the inner surface, while in Embolomeri it is wrapped around the lower surface so as to appear on the outer surface of the jaw. One other difference is of greater importance because, so far as is at present known, it distinguishes *Seymouria* from all fossil amphibia. In *Seymouria* the articular bone, by which the lower jaw is hinged to the skull is an independent ossification developed in cartilage, while in amphibia the corresponding joint is made by a mass which seems to be part and parcel of a membrane bone, the surangular of the outer surface of the jaw. The difference seems unimportant—certainly it has no mechanical significance—but it appears to be quite constant, allow-ing us to separate all reptiles from all amphibia.

Much more striking differences are to be found in the vertebral column. That of the Embolomeri (Fig. 46) is built up of individual vertebrae, each of which is in adults composed of three bones, two discs (the pleurocentrum and intercentrum) with concave ends perforated by a median hole, and a neural arch, which surrounds a canal for the spinal cord and rests on the two ventral discs, chiefly on the pleurocentrum but also on the intercentrum. In order to pre-vent rotation of the vertebrae on one another and to control the flexibility of the whole column the neural arch bears two pairs of processes, the zygapophyses, each having a smooth face by which it articulates with the zygapophysis of the neighboring vertebra. In all Embolomeri the zygapophyses are narrow and their articulating faces are at about 45 degrees to the horizontal, a position which allows

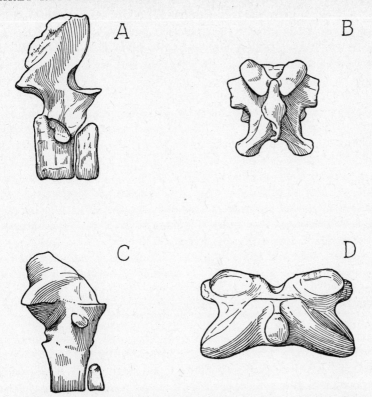

FIGURE 46. A dorsal vertebra of the embolomerous labyrinthodont *Cricotus* from the Lower Permian of Texas, compared with a dorsal vertebra of *Seymouria,* also from the Texan Permian.

A. Right lateral surface, *Cricotus.*
C. Right lateral surface, *Seymouria.*
B. Dorsal aspect, anterior end forward, of *Cricotus.*
D. Dorsal aspect of *Seymouria.*
To show the characteristic swollen neural arch with horizontal articular surfaces of the zygapophyses, characteristic of the basal reptilian group Cotylosauria. The *Cricotus* vertebra was chosen as the widest neural arch known among labyrinthodonts. Original.

the back to be moved up and down as well as from side to side. Finally there is a tall neural spine designed to give attachment to muscles which pass from vertebra to vertebra along the back.

In *Seymouria* the pleurocentrum has become much smaller in diameter and longer; it is cylindrical, the depression in its end and the perforation making it dicebox shape. The intercentrum is much smaller than the pleurocentrum and is no longer circular because its upper half has vanished, so that the bone is a crescent. The neural arch rests entirely on the pleurocentrum and is immensely heavy and wide. The zygapophyses stand out laterally and their articular surfaces are placed exactly horizontally. The whole thing looks bulbous and blown up; it is in part composed of a spongy bone with great spaces for marrow, covered of course by a thin, superficial layer of dense bone. The neural spine is never more than a small knob and may be entirely absent.

These structural differences imply corresponding changes in flexibility and in the distribution of the muscles. Comparison with snakes, which make some small approach to the same structure, and a consideration of possible movements show that the vertebral column must have been quite incapable of dorsoventral movement, while its lateral flexure can only have been slight and very powerfully controlled. The embolomerous vertebra shown in Fig. 46 has the widest neural arch of any I know, while the *Seymouria* vertebra with which it is compared is not so exaggerated in its development as others farther back in the same individual. The *Seymouria* vertebra is an extreme example of a condition found in nearly all members of the most primitive order of reptiles, the Cotylosauria. Relics of a similar structure are found, especially in the neck, in the primitive members of other reptilian orders, and there is no doubt that such features are typically reptilian and quite unlike any amphibian whatsoever. It is evident that the differences in the vertebrae imply corresponding changes in the musculature, not in all probability in the number but in the relative sizes and mechanical significance of the individual muscles.

The modification of the structure of the first two vertebrae, which are associated with the attachment of the skull, is of interest for our present purpose. According to an old description by Cope which there is no reason to doubt, the occipital condyle of an embolomerous skull articulates directly with a circular intercentrum exactly like all others in the animal, and there is no other contact between skull and neck. In *Seymouria* the knoblike condyle rests in a cup, the lower half of which is an intercentrum, the upper built up by two halves of the neural arch (Fig. 47). The ring formed by this group of bones rests on a special structure, the odontoid, which has a trefoil-shaped knob facing forward. The odontoid is a pleurocentrum whose whole character is changed for its specific purpose. The *Seymouria* structure is quite characteristically reptilian and can be matched by no amphibian.

In the pectoral girdle we have two characters found in all reptiles but in no known Carboniferous amphibian. The cleithrum, a small membrane bone lying along the front edge of the scapula, is lost in *Seymouria,* and the original cartilaginous shoulder girdle ossifies as two separate bones.

Only one Paleozoic amphibian, *Diplocaulus,* shows a similar arrangement, and this animal, a lepospondyl, can be shown to have developed the condition secondarily, so that the single casual exception is of no real importance.

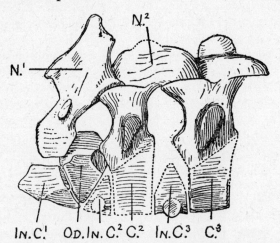

FIGURE 47. The first three vertebrae of *Seymouria,* to show that the first, the atlas, in its possession of an odontoid bone and all its other features, is characteristically reptilian in contrast to amphibian. C^2, C^3, centra of 2d and 3rd vertebrae; *In.C.*$^{1, 2, 3}$, intercentra 1, 2, and 3; *O.,* odontoid; N^1 and N^2, neural arches of 1st and 2d vertebrae. After Watson.

The humerus of *Seymouria,* unlike that of any known amphibian (except *Diplocaulus*), has a perforation near its lower end, the entepicondylar foramen, which exists to transmit a nerve and thus protect it from the pressure of the muscles.

Finally the fingers of *Seymouria* seem to have respectively two, three, four, five, and four phalanges or joints, while all amphibia have fewer. The *Seymouria* number is that found in all primitive reptiles.

The short series of characters listed above seem to be in no way linked to one another. They cannot be accounted for on mechanical grounds as the necessary sequels of one or few basal changes, and to my mind they imply that *Seymouria* is a reptile, or at least that it has progressed too far along the line leading to reptiles ever to have re-traced its steps.

The late Professor Sushkin, whose untimely death removed a man from whom much might have been expected, attempted to show that there were certain features of *Seymouria* in which it had advanced above the Embolomeri, which prevented it having any special rela-tionship to the reptiles. Professor Romer has dealt with Sushkin's evidence, and in a lucid discussion has shown that much of it rests on an insecure basis because the facts are incorrect and that the remainder is at least inconclusive.

But fortunately we can go further. *Seymouria* no longer stands alone in its group; we have materials of *Solenodonsaurus* from the Upper Carboniferous of Nyran, *Discosaurus* from the Lower Per-mian of Dresden, and *Kotlassia* and *Karpinskiosaurus* from the Upper Permian of Russia, all unquestionably allied. Although much of this material has been inadequately described, enough is known about all these animals to show that the group presents no obvious evolu-tionary changes paralleling those of the labyrinthodonts. As we have seen that the amphibian evolutionary trends are common to all its groups, we have now very strong evidence in support of the rep-tilian nature of *Seymouria*.

Dr. T. E. White, of Harvard University, has, however, discussed with and shown to me a fact about *Seymouria* which throws an en-tirely new light on its systematic position. He has found, in what are apparently full-grown skulls of a new small species, well-defined grooves for lateral-line sense organs. I regard this as conclusive evi-dence that some seymourias—not all specimens show the condition —were largely aquatic. But we can go much further. In the history of every amphibian, recent or fossil, in which the facts are known the larva, after it hatches from the egg and begins its life in water, develops lateral-line organs. These remain until it leaves the water at metamorphosis, being retained permanently in such forms as have an aquatic adult, even for example, in the frog *Xenopus*. Now the reptile egg must necessarily be laid on land, and the animal which hatches from it grows without metamorphosis into the adult. In no known aquatic reptile are lateral-line organs formed, presumably because during the development period the embryo is still within the egg or on dry land. It thus seems almost certain that *Seymouria* must have had a larval life in water and that it was, from the very important standpoint of its life history, an amphibian.

Nonetheless, although the well-known members of the group are too specialized in the otic region to have been ancestral to any other reptiles, and although even the earliest member of the group is much too late in time to be ancestral to them, it is, I think, legitimate to hold that apart from their obvious specialization the Seymouridae do preserve the structure of the earliest reptile stock, as it was before it left the amphibia to cross the boundary between the two groups.

It would therefore follow that the morphological changes which distinguish the skeleton of a reptile from that of an amphibian were gradually built up by the accumulation within a single animal of many of a series of minute modifications, each complete in the sense that the individual character is either amphibian or reptilian, the sum of all showing a gradual transition from one to the other structure.

But the essential change lies in the establishment of new individual evolutionary trends, which, followed out in the later history of the group, control the fundamental structural and other changes.

It is interesting, and it is necessary for my purpose, to consider in brief outline the early history of the reptiles. In the Basal Permian rocks of Texas and New Mexico are found large and magnificently preserved faunas of reptiles, the earliest known in the world. Fragmentary remains of similar or slightly later age from Europe show that these faunas are representative of their time and that we are justified in building on them our conception of the early development of the group.

The fact that, like the Paleozoic amphibia, the seymouriamorphs have skulls in which the whole of the exterior was covered with a continuous mosaic of dermal bone, perforated only by the orbits, nostrils, and a foramen for the pineal gland (still presumably a functional eye), shows that we are justified in considering the order Cotylosauria, in the members of which this quality is still retained, as fundamental. It is a primitive group, or at least a group retaining primitive features lost by all other reptiles.

Seymouria inherited from its amphibian ancestors an otic notch, a deep incision in the side of the skull, open behind and stretching forward toward the eye. In the dried skull this leads down into a space, the tympanic cavity, which completely separates the side of the ear capsule from the pterygoid, the quadrate, and the inturned part of the squamosal which wraps round the hinder surface of this bone. The tympanic notch was covered in life by a thin, stretched membrane into which was inserted the outer end of the stapes which, crossing the tympanic cavity, transmitted its vibrations to the inner ear. Consideration of the comparative anatomy of the stapes in recent reptiles and of the conditions described by Sushkin in the labyrinthodont *Dwinasaurus* shows that the outer end of the stapes was slung up to the tabular bone by a special process, and that a descending cartilaginous process passed down from it to the base of the tongue. The primitive cotylosaur reptiles show one or other of two possible modifications from this condition. From the nature of these it is clear that once committed to one or other no return was possible. I have therefore used the conditions here as the basis of the classification of these reptiles. Indeed it is possible and desirable to go further and divide the whole of the reptiles into two branches in the same way.

It appears that for some reason, probably merely mechanical, it became desirable to place the quadrate bone vertically. It is evident that if we begin with *Seymouria* this can be done in two ways, or rather that only two ways seem reasonable, the conceivable hybrid conditions presenting no advantages.

These ways are as follows:

In one the upper end of the quadrate remains fixed and the lower end is swung forward. By this process the otic notch is preserved and indeed widened out. The upper end of the quadrate and all the bones associated with it retain their old wide separation from the end of the paroccipital process, which projects from the ear capsule to the tabular; and the end of the stapes, with all its processes, remains as it was. This condition is found in *Diadectes* (see Fig. 48) and is preserved in many later reptilian orders, including all those which have still-living members. This group ends in the birds.

The other way in which the quadrate can become vertical is by keeping the lower end fixed and moving the upper end backward. By such a movement the upper end of the quadrate and the bones associated with it are brought into direct contact with the end of the paroccipital process. The otic notch is entirely obliterated and the tympanic membrane, if it were to survive at all, would have to change its position. These modifications eventually change the whole arrangement at the outer end of the stapes. The disappearance of the tympanic membrane abolishes the possibility of a transmission of

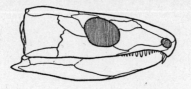

CAPTORHINUS
L.Perm.

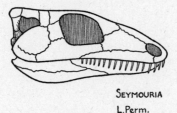

SEYMOURIA
L.Perm.

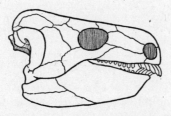

DIADECTES
L.Perm.

FIGURE 48. The skulls and lower jaws of three cotylosaurian reptiles, to show the two mutually exclusive ways in which the quadrate, to which the lower jaw is articulated, may become vertical. *Seymouria* retains the primitive conditions; in *Captorhinus* the upper end of the quadrate moves backward so that the otic notch, spanned by the tympanic membrane, is obliterated; in *Diadectes* the lower end of the quadrate swings forward and the otic notch is widened. All the mammal-like reptiles follow the *Captorhinus* type, all other reptiles agree in respect to this quality with *Diadectes*. *Diadectes* after Romer, the others original.

117

its vibrations through the stapes to the ear, and the animal would become deaf unless an alternative method could be devised.

It is a familiar fact that in men who are deaf through some defect of the tympanic membrane or of the sound-transmitting apparatus, but in whom the receptive organs and nerves are intact, hearing can take place by direct transmission of sound vibrations through the bones of the skull. Among living reptiles the snakes have no tympanic membrane but still hear, doing so because in them the end of the stapes is connected with the quadrate bone which, having a large area and lying immediately below the skin, can be thrown into vibration by sound waves and thus act as a tympanum. The extinct reptiles like *Captorhinus* belonging to my second group adopted the same device; in them the stapes gains a new attachment to the quadrate, and even in the absence of a tympanic membrane in adults they no doubt heard perfectly well.

To this second group of the reptiles belong those with which I shall deal in the remainder of this and the next chapter. None of them survived the Trias, but from them arose the mammals, and in the details of their evolution, now very well known, we find a complete explanation of the structural peculiarities of their descendants.

The existence of a contact between the head of the quadrate and the end of the paroccipital process, which in early forms of the group were separated by a thin layer of squamosal, has very important consequences. In *Seymouria,* as in the labyrinthodonts, all the muscles of the dorsal part of the neck were concentrated on the relatively small triangular area which is the back of the brain case. It is clearly desirable that this area should be increased, and under certain circumstances it would be advantageous to have some laterally placed neck muscle insertion widely separated from its fellow and placed low in the skull. These improvements of the structure will be secured if the outer ends of the paroccipital process move downward. In mammal-like reptiles they did so. This movement can only take place by a downward migration of the paroccipital along the inner side of the quadrate, and this process goes so far that the articulation between them comes to lie immediately above the condyle for the attachment of the lower jaw. The quadrate then secures so adequate a buttressing that the quadrate ramus of the pterygoid comes to have very little functional importance. In moving down to this extremely ventral position the paroccipital automatically drives the stapes before it, and this bone comes to articulate with the quadrate only just above its articular margin. This brings us to the position from which we shall start in the next chapter.

The first stage in the evolution of the mammal-like reptiles is the development of a hole in the originally continuous bony cheek. This opening, covered of course by skin, exists in them in order that the great muscle which moves the jaw may have room to expand in thickness when it shortens, as all muscles must do. Its appearance is probably to be correlated with changes in the direction and an increase in the size and power of these muscles, in a way first suggested by L. Dollo, but placed on a sound basis by W. K. Gregory. In this way we arrive at an animal like *Mycterosaurus* (Fig. 49) or

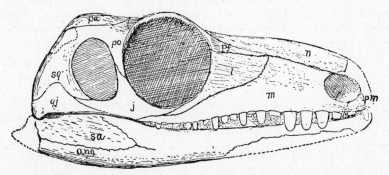

FIGURE 49. The skull of *Mycterosaurus,* perhaps the most primitive known mammal-like reptile. The small temporal fossa behind the orbit is characteristic of the group. *Ang.,* angular; *j.,* jugal; *l.,* lacrimal; *m.,* maxilla; *n.,* nasal; *pa.,* parietal; *pf.,* prefrontal; *pm.,* premaxilla; *po.,* postorbital; *qj.,* quadratojugal; *sa.,* surangular; *sq.,* squamosal. After Williston.

Varanosaurus (Fig. 50) from the Texan Permian. The general build of such animals was first described by Broili, but Williston, with much better materials, added greatly to our knowledge. *Varanops* (Fig. 51) is an animal built almost exactly like such a lizard as *Varanus,* differing only in the absence of a neck. It has a long, slender head, square cut in section, with laterally placed eyes. The mouth is long and is bordered throughout by numerous small teeth of which those in the premaxilla turn backward so as to make a prehensile hook, while two lying side by side a little way behind the nostril are enlarged. The back is long, there being twenty-seven vertebrae in front of the hind legs. The legs are spraddling, with the feet far apart, indicating a short stride, but have powerful and well-finished bones; both hands and feet have long digits ending in sharp claws. There is

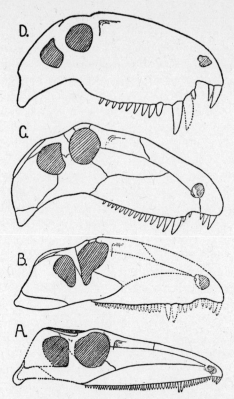

FIGURE 50. A series of drawings of the skulls of four of the pelycosaurs, primitive mammal-like reptiles, to show the gradual specialization of the dentition and the increased depth of the skull necessary to accommodate the long roots of the enlarged canine teeth. A, *Varanosaurus;* B, *Ophiacodon;* C, *Sphenacodon;* D, *Dimetrodon.* From Watson.

FIGURE 51. Skeleton of the primitive mammal-like reptile *Varanops,* to show the spraddling, lizardlike build, with widely separated plantigrade feet. After Williston.

no doubt that the creature was active and capable of quick movements.

From such an animal we can trace a morphological series of contemporary forms leading to the great and very spectacular *Dimetrodon* (see Fig. 50). The changes in the skull can all be related to an increasing differentiation of the teeth, associated with the adoption of more strictly carnivorous habits. The first change is an increase in the size of the premaxillary teeth, the anterior ones becoming bigger than the rest, and a corresponding increase in the length of the two enlarged teeth seen in the maxilla of *Varanosaurus*. This process goes on until the three remaining incisors become a series graded from front to back, and the canine teeth near the front of the maxilla become much larger than both those lying behind them and the one or two tiny ones which lie in front. Thus we arrive at a stage in which two relatively huge teeth are separated from one another by an interval occupied only by small teeth.

This condition depends on the fact that a single corresponding canine tooth in the lower jaw bites into the notch, passing up into the skull through a hole confluent with the internal nostril. This enlarged lower-jaw tooth cannot develop at the extreme anterior end because the bony girder occupying the middle of the roof of the mouth is mechanically necessary and cannot be sufficiently cut into to make space for a lower tusk near to the middle line. Great tusks like those of *Dimetrodon* have necessarily very long roots, and in order to provide a cavity in which a replacement tooth can be formed it is necessary to have a deep and thick maxilla. This can only be provided by carrying the lower tooth-bearing border of the bone downward, and the result of so doing is to produce a deep step in the upper jaw, the lower jaw acquiring a similar arrangement.

But these changes involve others; the palate can no longer stay flat, as it is in *Varanosaurus*, but becomes vaulted, because the bones in its mid-line are connected in front to the premaxillae and behind to the brain case and thus remain fixed when the maxillae grow downward. The vaulted palate has many advantages; it has, automatically in that it is a curved sheet, greater strength than a flat palate, and it reduces the danger of the animal being unable to breathe when its mouth is filled with a struggling, just-captured prey.

Dimetrodon is a large animal, about six feet in length, and is remarkable because the neural spines of all vertebrae between skull and hind legs are produced into tall, upstanding rods of bone about two feet high, the whole structure, tied together by a tough envelope of skin, standing up as a crest along the middle line of the back. It is a specialization, with no obvious function, of a kind very often seen in the last members of a group, animals which so completely dominated their environment that they could afford to produce ornamental structure even if it reduced their general efficiency. In the general character of its limbs *Dimetrodon* shows very little advance over *Varanosaurus*, but a small alteration in the articulation between the arm and shoulder seems to show that the stride was proportionately longer.

The mammal-like reptiles of Middle Permian times are known only from specimens found long ago in the copper mines of the Ural Mountains. Although the existing descriptions of these important fossils are inadequate, it is at least clear that they do represent an intermediate stage between such animals as *Varanosaurus* and the most primitive mammal-like reptiles of Africa.

South Africa is unique in that it contains beds which, from the time of the Upper Permian to the Upper Trias, give us a continuous record of a land fauna. Animals are known of all sizes, from a rat to a rhinoceros, often perfectly preserved, and of all ages. We have in fact in South Africa the only case known in which it is possible to study the evolution of groups of reptiles in as much detail as in the case of Tertiary mammals. It is already possible to sort out from the hundreds of species of mammal-like reptiles from the Karroo a number of approximate evolutionary series (one such forms the basis of the next chapter), and a general study of the whole material allows us to obtain a clear conception of the general course of evolution within the group.

One evolutionary line runs through a group called the Gorgonopsia to the Cynodontia, but the actual line of descent cannot yet be disentangled from many others which, for a greater or less period,

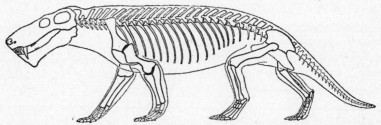

FIGURE 52. The skeleton of the gorgonopsid *Scymnognathus*. This is an advanced mammal-like reptile, with a mammal-like pose and mode of walking. The hands and feet are digitigrade and not far apart. From Broili.

pursue parallel evolutionary courses. Direct comparison of a gorgonopsid with a Lower Permian pelycosaur is difficult but useful.
If we consider the skull first we find that the differences can be accounted for in part only on mechanical grounds. The nature of their
dentition shows that the gorgonopsids were carnivores which fed on
other large animals; thus they needed enlarged canine teeth for
killing their prey and powerful incisor teeth for tearing flesh from
bones. By analogy with recent carnivores they would require an elaborated and specialized series of teeth behind the canines, the anterior
of these serving to crush while the posterior sectorial teeth would
work like scissors, cutting up muscle into easily digestible pieces.
The incisors of early Gorgonopsia are numerous, powerful, and often
elongated teeth arranged around a semicircular muzzle; they are
opposed by similar teeth in the lower jaw which bite within them.
Behind the incisors, exactly as in *Dimetrodon* and for the same reason, there is a gap. Behind the gap the upper jaw suddenly steps
downward and bears a canine tooth, which is sometimes a huge,
flattened dagger like that of a saber-toothed tiger. The resemblance
of this part of the jaw to that of *Dimetrodon* is very considerable.
There is no doubt that the arrangement has arisen independently in
the two groups as the easiest way for allied stocks, arising together
from some common ancestor, to meet the same demands.

From this point backward the teeth of the two groups differ widely.
Dimetrodon has a long series of some ten or a dozen teeth, a number
which may be brought up to sixteen when alternate sets of teeth are,
transiently, present together, while the gorgonopsids generally have
only four or five cheek teeth. I imagine this difference means that at
a time when all teeth were simple cones a long series was useless
in a carnivore whose canines provided an effective mode of killing
prey and incisors a method of tearing it up.

The very powerful dentition of gorgonopsids could only be used
successfully if the animal possessed very powerful jaw muscles. Those
of gorgonopsids are huge, apparently as great in their transverse sectional area as in any mammal of the same size.

The final term of the evolutionary series, a cynodont, is an animal
with a large head and a long but very powerful body, mammal-like
in the possession of seven vertebrae in its neck and in a distinction
between a thorax with long movable ribs and a lumbar region in
which the ribs are immobile and indeed articulate with one another.
The limb girdles are very mammal-like. The limbs, like those of the
more primitive mammals, are held with the elbows and knees some-

what everted and with the forearm and lower leg at an open angle
with the proximal bones. The feet were digitigrade, neither the palm
of the hand nor the sole of the foot resting on the ground, and the
ankle joint was very mammalian, there being a heel and special facets
for locking the two proximal bones firmly to one another.

The cynodonts had a long series of cheek teeth, differentiated into
simple pegs behind the canines and cutting or grinding teeth pos-

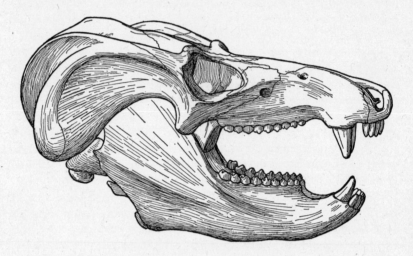

FIGRUE 53. The skull and lower jaw of *Gomphognathus* to show its thoroughly mammalian appearance and the nature of the jaw articulation. From
a model by Mr. F. O. Barlow of the British Museum.

teriorly. It is evident that the full usage of a grinding series of cheek
teeth, such as occurs in some cynodonts, is impossible or unlikely in
an animal in which the palatal nostrils open far forward on the roof
of the mouth, and the inspired air passes backward through a food-
filled buccal cavity. Even the presence of a median groove on the
palate, though certainly helpful, cannot have made the whole arrangement satisfactory; therefore the width of the opening into the
groove became narrowed by the ingrowth of special folds of skin
from the sides which came to be supported in the mid-line by a thin,
bony plate growing down from the bottom of the groove to meet
them. Finally the whole arrangement is made solid by the ingrowth
into these folds of skin of special bony secondary plates of the
maxilla and palatines. The structure so formed is the cynodont (or
mammalian) secondary palate.

It is most probable that cynodonts had already developed a mammal-like muscular cheek which, in association with the tongue, could hold food in position between the teeth and drive it to and fro. Such a muscular cheek implies a muscular skin of a soft structure unlike the scale-covered skin of reptiles in general. The presence in cynodonts and other advanced theriodonts of very large foramina, through which branches of the fifth cranial nerve came out onto the face, suggests that the skin of the snout contained many sense organs and that there was an area homologous with the rhinarium, the naked area of skin surrounding the nostrils of mammals.

There is a deep groove in the skull of cynodonts which certainly housed a long, tubular, external auditory meatus leading down, as in a mammal, to the ear drum; and I should not be surprised if there was a simple immobile pinna or external ear. The internal ear is shown by the cavity which housed it to have possessed a short, curved cochlea, so that it is probable the animal had an analytic apparatus for sounds similar to that of a mammal.

Perhaps even more interesting is the evidence of the structure of the nose. Mammals, perhaps because they are warm blooded, have the first portion of their nasal cavity subdivided into a series of very narrow chinks by folds of mucous membrane, which are supported by curious, very delicate bony lamellae, the naso- and maxillary turbinals. These bones are attached to the bony side walls and roof of the nasal cavity but do not usually fuse with them. In many more primitive mammals the attachment of the turbinals is to low ridges on the inner surface of the bones of the face. In some cynodonts these ridges are to be found in the same position and are of the same character as in mammals and it is reasonable to assume that they had the same function of supporting turbinals. The skin covering the naso- and maxillary turbinals, although sensory, has nothing whatever to do with smell; it determines the temperature, wetness, and so on, of the inspired air. The whole arrangement of these structures is such as to provide an air-conditioning plant which will insure that the lungs receive only warmed, humidified, and as far as possible, dustless air. The sense of smell is exercised by the skin covering another set of turbinal bones, the ethmoturbinals, which lie in a side chamber above and behind the main air passage. They occur in cynodonts as in mammals. Thus the development of the naso- and maxilloturbinals in cynodonts suggests that, like mammals, they were warm blooded.

APPENDIX

The problem of the origin of the reptiles has been discussed extensively since the fifth lecture was delivered, and it is necessary to pay attention to some of the evidence and conclusions which have lately been published. It may, I think, be said that nearly all authors who have recently written on the matter have accepted, though sometimes in a modified form, the view that all reptiles have passed through a stage in which an ancestor, if known, would be placed in the group called Cotylosauria. And most or all of these authors have accepted the view that the division of Cotylosauria into Captorhinomorpha and Diadectomorpha is justified, although they hold different opinions as to the origin and relationship of these two groups to one another. There is, I think, a wide agreement that the anthracosaurs include the ancestors of the cotylosaurs, but Westoll demurs to this view, bringing forward arguments to show that the captorhinomorphs, at any rate, which are known back in Stephanian times, may have been derived from such adelospondyls as *Microbrachis* and *Hyloplesion.* If this view were true it would follow inevitably that *Seymouria* and its allies might conceivably be related to the diadectomorphs, but could not be close to captorhinomorph ancestors. On the other hand Olsen, with the concurrence of Westoll, finds a relatively close correspondence between *Seymouria* and *Diadectes* itself. It seems desirable to discuss these views shortly and perhaps in a somewhat abbreviated manner.

On the amphibian side the relationship of the seymouriamorphs seems perfectly evident. The pattern of the dermal skull roof is essentially identical with that of the anthracosaurs and is found nowhere else, and as I have shown in the chapter the rest of the skeleton shows nothing inconsistent with such relationship. But it is evident that the seymouriamorphs show a number of advances in structure above the conditions found in the Coal Measure and Namurian anthracosaurs. The changes are, in general, parallel to those which are found in labyrinthodonts. The brain case becomes less well ossified, in all of them the sphenethmoid ceases to have any contact with the pro-otic, and indeed it is probable that the large gap between these two bones was spanned by a membrane and not a cartilage. In *Seymouria* the reduction goes so far that the pro-otic loses all contact with the roof of the skull, although in the later *Kotlassia* it still has a long contact with the supratemporal. In both *Seymouria* and *Kotlassia* the postparietals and tabulars have developed occipital flanges, which pass down on the posterior surface of the neural cranium exactly as in labyrinthodonts, and the posterior temporal fossa is enlarged. No important modifications take place in the palate or in the general proportions of the skull; but there is a tendency, just indicated in *Seymouria* but well exhibited in *Kotlassia*, for the lower part of the quadrate to become vertically placed, and its articular surface lies very nearly on the same plane as the occipital condyle in each of these forms, a point of marked divergence from usual anthracosaur structure. Although the skull of *Kotlassia* is less high than that of *Seymouria* it is still higher than in the majority of later labyrinthodonts. The lower jaw of both forms is typically labyrinthodont, differing from that of anthracosaurs in general in the reduction of the two enormous internal vacuities to rather small foramina. *Seymouria* has no retroarticular process; in *Kotlassia* on the other hand there is a small one. It may be recalled that the anthracosaurs completely lack any such structure.

In the vertebral column one of the most remarkable features of *Seymouria*, which appears to be repeated in essence by *Kotlassia*, is the character of the atlantoaxial complex, which in *Seymouria* certainly, and in *Kotlassia* with great probability, exhibits a typical reptilian odontoid, articulating by separate facets with the paired neural arches and the intercentra, the single occipital condyle fitting into a cup on the neural arch and intercentrum. The vertebral column of *Seymouria* could well be derived from the embolomerous condition by a relative elongation of the ring-shaped pleurocentrum and a failure of the dorsal part of a ring-shaped intercentrum to ossify. *Kotlassia* differs in a great further elongation of the pleurocentral ossification. Both forms differ in the heavy, widened neural arch with swollen posterior zygopophyses; these structures are far better developed in *Seymouria* than in the later *Kotlassia*.

White has shown that a reduced cleithrum occurs in the shoulder girdle of *Seymouria*, and *Kotlassia* has the same bone. In *Seymouria* the shoulder girdle contains both a scapula and a coracoid with a characteristic retention of a screw-shaped glenoid cavity. In *Kotlassia* the coracoidal end of the shoulder girdle is fully ossified, but it is uncertain whether it is separate from the scapula or not. The short limbs of *Seymouria* are extremely powerfully developed and do not differ fundamentally from those of *Eryops,* but the elongated *Kotlassia* has very slender limbs, fundamentally no doubt of the same type but very different in general appearance.

Additional evidence about seymouriamorphs is perhaps provided by that remarkable animal, *Lanthanosuchus,* which comes from the second of the Russian Permian zones and is thus intermediate in date between *Seymouria* and *Kotlassia.* This animal, known only by a single skull, is one of the most extraordinary creatures yet found. The skull, which I have seen, is absolutely perfectly preserved; it looks indeed like fresh bone. It is quite large, about 20.75 cm in width and 17.7 cm in length, but its maximum height at the occiput is 4.5 cm. Anteriorly the skull rapidly shallows and becomes very low. The upper and lower surfaces are therefore essentially flat. In the palate the admesial margins of the pterygoids are in contact with the parasphenoid for a very considerable distance, and the parasphenoid is short. The palate is closed, the internal nostrils large, and the quadrate condyles lie just behind the occipital condyle. There is a completely ossified occipital ring, basioccipital, exoccipital, and supraoccipitals. The paroccipital is well ossified on its posterior surface, but there is no ossified pro-otic at all; nor, I think, is the basisphenoid ossified. The epipterygoid is a solid column connecting the pterygoid with the skull roof. The pattern of the dermal roof is unique, the parietals extend backward so as completely to separate the postparietals. The tabulars and the posterior part of the squamosals extend backward as a great shelf far behind the vertical part of the pterygoid and quadrate. There is no supratemporal, the tabular directly meeting the postorbital. The lachrymal does not approach the nostril. The irregularly shaped orbit faces directly upward and the eyes must have protruded above the upper surface of the skull. But the most astonishing feature is the presence on each side of a single, large temporal vacuity bounded by the jugal, postorbital, squamosal, and quadratojugal. The significance of this opening is obvious; it lies above the transverse flange of the pterygoid at its junction with the ectopterygoid and must have existed to give the possibility of thickening to the masticatory muscles, which passed, dorsal to the palate in this region, down to the lower jaw. The occipital condyle, made by the basioccipital and exoccipitals, is thoroughly seymouriamorph, and the skull pattern and indeed the whole structure of the palate are readily derived from conditions in *Seymouria.* There is no other group of labyrinthodonts or of reptiles in which the genus can be placed, or from whose members *Lanthanosuchus* could have been derived, and I therefore regard Professor Efremov as correct in his conclusions that it came from more normal seymouriamorphs.

It should be mentioned that in the very extensive excavations in the Ishyevo locality, from which this skull came, there were found a few typical seymouriamorph vertebrae appropriate in size to this skull, showing the existence of an animal whose skull is not represented by a single fragment, unless indeed it be *Lanthanosuchus*. The interest of the genus lies in its extreme flattening. It would indicate that even seymouriamorphs exhibit the characteristic labyrinthodont phenomena of dorsoventral compression and reduction of cartilage bone. Thus the seymouriamorphs are undoubtedly to be regarded as amphibia, but the reptilian qualities in their skeleton still remain, and taken together they are so striking as to provide strong evidence that the group is really very closely related to the reptilian stock.

Many years ago I showed that the Cotylosauria, the most primitive reptilian group, is, after *Seymouria* has been excluded from it, divisible into two groups, the Diadectomorpha and the Captorhinomorpha, which are distinguished primarily by the retention of an otic notch in the first group and its loss in the second. I was in difficulty with regard to one form, the very remarkable reptile *Limnoscelis*. Recently Romer, in a new account of *Limnoscelis*, has suggested that this animal, which is clearly of very archaic type, represents a form almost immediately derived from amphibia in which the otic notch has been obliterated essentially by a backward growth of the supratemporal along the ventral margin of the tabular and an upgrowth of the squamosal, so that ultimately its suture with the temporal extends to the extreme posterior end of the tabular and the occiput is in effect continuous with the posterior surface of the squamosal, where it turns around behind the quadrate, whose posterior margin must hence stand almost vertically. Romer then goes on to suggest that from such a condition the otic notch of *Diadectes* for example may be redeveloped essentially by a movement forward of the now vertical quadrate as a whole, leaving behind in evidence a contact between the head of the quadrate and the outer surface of the brain case probably on the pro-otic. Romer suggests—without I think completely establishing the point, though he is in all probability right—that the fenestra ovalis of *Limnoscelis* lies very ventrally on or below the level of the floor of the brain case, a little laterally displaced, at the end of a short, horizontally directed process which descends abruptly from the lower surface of the paroccipital. Unfortunately the stapes is not preserved. I have examined the only known skull of *Limnoscelis* on several occasions and have never been able to satisfy myself as to the real structure of the occiput, but it may well be essentially as Romer describes it. It will follow therefore merely as a matter of definition that *Limnoscelis* is a captorhinomorph, as indeed I suggested in 1917. If so it is vastly more primitive than any other, and its structure becomes of very great interest because of its bearing on the origin of reptiles in general. As I understand it, Romer would not object to this view.

It would follow, therefore, that the assumption that the closure of the otic notch in captorhinomorphs was associated with the achievement of a vertically standing quadrate by the moving backward of the upper end of that bone may well be justified, and it is desirable to consider the conditions in diadectomorphs in order to see whether they are secondarily derived from such a condition as that found in *Limnoscelis* or arose by a forward swing of the lower end of the quadrates, leading, as I originally suggested, to a retention, and even an enlargement, of the original amphibian tympanic membrane.

An important paper by Olson bears directly on this matter. Olson claims to find in *Diadectes* the retention of a seymouria-like arrangement of tabular, supratemporal, and intertemporal, all articulating laterally with the parietal. I am not convinced that this structure actually exists in these forms. It is certain that his drawing of the dorsal surface of *Diadectes* cannot be altogether correct, for in his interpretation he makes the sutures between the intertemporal and the parietal run continuously forward into that between the postfrontal and the frontal, and it is quite certain from the figures given by Broom, and from the magnificently preserved type skull of *Nothodon*, that this condition does not occur; the bone called intertemporal by Olson (and squamosal by Williston) ends abruptly in contact with the posterior margin of the parietal, which extends outward laterally to the lateral margin of the supratemporal. The specimen in the American Museum of Natural History (No. 4378, described and illustrated by Broom in 1914) is perfectly preserved and shows the whole structure of this region exceedingly well, agreeing to my knowledge with Broom's figure; Dr. Colbert has reexamined it on my behalf and tells me that the figure is correct. The separation of a supratemporal from an intertemporal therefore appears not to exist. Thus the apparent close resemblance to *Seymouria* is, I am afraid, not to be found.

Olson however draws attention to another feature of *Diadectes* which does appear to present a real resemblance to *Seymouria*. This is the remarkable way in which in *Diadectes* the fenestra ovalis is carried out laterally and considerably posteriorly, so that it lies far out toward the end of the rather thin and highly peculiar distal extremity of the paroccipital, here fused indistinguishably with the tabular. The fenestra ovalis is connected with the ear by a long tube which lies on the lower surface of the fused paroccipital and pro-otic. These bones were obviously each continued by a cartilage, the two together presumably completely surrounding the canal; and the ventral surface of the entire structure is, as Olson has pointed out, sheathed by a remarkable, laterally produced process of the parasphenoid, which must have reached out to, or very nearly to, the actual fenestra. This structure does really recall that of *Seymouria* where, as White has shown (following Sushkin), the fenestra ovalis lies exceedingly far laterally and the parasphenoid extends out below the joint pro-otic and paroccipital, so that it forms a small part of the ventral border of the fenestra, a retention, in a very remarkable form, of normal labyrinthodont conditions. Bystrow has confirmed the view of Sushkin that the fenestra ovalis of *Kotlassia* occupies a similar position, and from examination of actual materials I am inclined to believe that in this animal, as in *Seymouria*, the parasphenoid does actually extend out to the fenestra. This resemblance between *Diadectes* and *Seymouria* seems to be genuine, but its significance is difficult to determine. It is a specialized character of seymouriamorphs which does not occur in other labyrinthodonts, and I find difficulty in imagining that it can have arisen except from a stage of the general nature of that found in *Edops* or indeed in many other primitive labyrinthodonts. What the functional significance of the whole may be I do not know; it has the result of shortening and lightening the stapes, which may have conduced to a more satisfactory transmission of sound waves from the tympanic membrane to the perilymph. But this structure occurs in no other cotylosaur; it seems to be a special peculiarity of *Diadectes* and may have arisen independently.

The other diadectomorphs, the pariasaurs and procolophonids, are very much later in date. The earliest of them is an undescribed reptile from Belebei (Bashkiria) in the second zone of the Russian Permian succession, and only a little later are the *Nyctiphruretus* and *Nycteroleta* from the Mesen fauna. Then follow the pariasaurs and the more advanced procolophonids.

Nyctiphruretus and *Nycteroleta*, of which I have seen the material, have an extraordinary occiput in which the paroccipital process stretches upward in a typically labyrinthodont manner to articulate with the tabular, placed quite laterally in a labyrinthodont manner at the lateral end of the occiput, forming the upper border of the very large otic notch. The stapes is known as a fairly heavy bone, which Efremov believes to have articulated with the quadrate, but I am uncertain whether this is in fact true. In *Procolophon* the stapes does not articulate with the quadrate; on the contrary, it passes out so that its distal end is directed toward the tympanic notch at about the most ventral point of the downturned tabular. It is too short to have reached a tympanic membrane directly, unless the membrane were sunk below the surface of the head, but could have been connected with it by an extracolumella.

It is an important point that the fenestra ovalis of *Procolophon,* the borders of which are not very well ossified, is in the bony skull, partly formed by a special lappet from the parasphenoid, or at any rate from a perichondral bone sheet projecting from the joint basi-parasphenoid, which is presumably actually parasphenoid. This may be a relic of the diadectid-seymouria condition. It is possible that a similar arrangement exists in *Nycteroleta* and in *Nyctiphruretus.* In *Procolophon* itself the lateral parts of the paroccipital are much deepened, the bone is horizontally directed, and the tabular is pulled well down onto the occipital surface, a rearrangement presumably associated with the development of the neck musculature. *Pariasaurus* shows what is essentially a similar condition, and I understand from Professor Efremov that it is not improbable that the pariasaurs and the procolophonids had a common origin, perhaps not earlier than Zone I of the Russian Permian. Indeed comparison of my figures of the occiputs of *Pariasaurus* and of *Procolophon* will show that such a view is perfectly practicable; it may be pointed out that the arrangement of the teeth on the palate of *Nyctiphruretus* much recalls that on the pariasaur palate as illustrated by Seeley.* It appears therefore not inconceivable that the Diadectomorpha is a group of genuinely related animals, but if this be so then clearly its members have not passed through the stage represented by *Limnoscelis,* for it seems quite impossible that the thoroughly labyrinthodont-like occiput of *Nyctiphruretus* could ever have been secondarily redeveloped from such a form, though it could quite easily have come from a seymouria-like form.

Many years ago I suggested that loxommids, whose stapes is known to pass from the side of the otic capsule outward and upward, being inserted into a tympanic membrane which crossed the otic notch,

* H. G. Seeley, *Phil. Trans.* (1892), B, pp. 311–370.

possessed a cartilaginous extension from the stapes passing down-ward to be inserted into a deep pit on the inner side of the quadrate just above the articular margin. *Seymouria* shows a boss of highly peculiar character on the posterior surface of the quadrate, and this boss may be an indication that a similar hyoidean string arose from the stapes, which is not very well preserved, and passed down-ward and backward to be connected with it. Bystrow has shown that a similar process occurs in *Kotlassia*, though here its character is rather different. Finally an identical condition has long been known in *Diadectes*, and it is reasonable to suppose that here also it has the same significance. The implication of these facts is that in all these animals there existed the homologue of an extracolumella, attached to the end of the stapes with the dorsal process slung up to the tabular on the back of the skull, and with a hyoid branch extending ven-trally, presumably to continue into the ceratohyal, the condition being in fact exactly as in *Sphenodon* and other still living reptiles. Captorhinomorph cotylosaurs are still poorly known; we have ade-quate materials of *Captorhinus* itself, less complete of its successor *Labidosaurus*, and of the other forms we know little. But two of them, described by Price as *Romeria* and *Protorothyris*, which are the earliest known members of the group, though still very incom-pletely known, are evidently clearly related to *Captorhinus* and equally nearly to the ancestry of the pelycosaurs.

The most characteristic feature of all these reptiles is the square-cut appearance of the posterior border of the skull, where a nearly continuous face extends, in *Captorhinus*, from the interparietal la-terally over to the squamosal and so right down about to the articular border of the quadrate. As Romer points out there is a very small projection, actually just below the lateral corner of the supratem-poral, which occurs in these forms. It is conceivable that this might mark the point of attachment of an actual tympanic membrane pass-ing downward to the upper surface of the retroarticular process of the lower jaw, which is quite well developed in this form. But ex-amination of actual specimens seems to show that this view is no more than theoretically possible, for the membrane so stretched would not be flat, would be exceedingly high in comparison with its width, and could not be approached by the end of the stapes. In a paper now in the press (Watson, 1948) I have shown that it is very improbable that the pelycosaurs, which derived from captorhino-morphs, possessed a tympanic membrane at all; and I think it highly probable that none existed in *Captorhinus*.

At any rate in *Captorhinus* the stapes is a massive rod with an enormously expanded proximal end which, at the fenestra ovalis, articulates with the basioccipital, the paroccipital, and pro-otic, and, over a very small area, with an extreme lateroposterior extension of the parasphenoid. The bone is rigidly connected with the otic capsule by a dorsal process articulating with the pro-otic. The distal end of the bone, which clearly possessed a cartilaginous extension, fits into a groove and pit on the inner and posterior aspect of the quadrate. Indeed in one skull (No. R. 9) in my own collection its distal end is co-ossified with the quadrate. It seems quite impossible that a bone of this character could have possessed a functional extracolumella part extending outward to a tympanic membrane. It is evident that the so-called dorsal process is not homologous with the dorsal process of the columella in living reptiles; probably it arose from the curious small mesially and generally forwardly directed process of the stapes, which occurs in many labyrinthodonts and is not represented in the stapes of any reptile not in the mammalian line of descent. The stapes is unfortunately not known in *Limnoscelis,* but from the general relations of the parts in that animal it seems probable that it conformed to the captorhinomorph rather than to the common reptilian condition, indeed, as there is no position in which a tympanic membrane could be attached to the *Limnoscelis* skull, it seems evident that the stapes would have had no need of an extracolumella.

Thus the magnitude of the gap which exists between the diadectomorphs and the captorhinid pelycosaur stock is very great, and the conclusion seems inescapable that all the living reptiles and those extinct groups which are definitely related to them, deinosaurs for instance, must have been derived from diadectomorph ancestors, while the mammals are clearly of captorhinomorph ancestry. It is at any rate highly probable that the cleft between the mammals and birds extends back to the very beginning of reptiles, and it is interesting to consider whether these two reptilian stocks arose independently from amphibia or not. This matter has been raised recently by Westoll, who suggests that the captorhinomorphs may have come from such Adelospondyli as *Microbrachis.* I think such ancestry is in fact impossible, but prefer not to discuss the matter, which is not necessary for my present purpose, on this occasion.

Romer has recently published an admirable account of the pelycosaurs as a whole, showing that *Varanosaurus* belongs to a stock separate from that from which *Dimetrodon* arose, and that *Dimetrodon,* or rather a closely related form, must have been ancestral to the later mammal-like reptiles. I have now had the advantage of seeing most of the material of those early mammal-like reptiles which has been found in the copper-bearing sandstones of the Urals and in somewhat later beds at Ishyevo on the central Volga.

These forms are of extraordinary interest; they fall into two groups, one ancestral to the tapinocephalid Deinocephalia of South Africa, the other to the titanosuchid Deinocephalia, and also I think, to the gorgonopsids. There is some evidence that the Therocephalia were actually in existence in Zone II of the Russian Permian. The two important reptiles from the copper-bearing sandstones of Zone I, *Rhopalodon* (which should actually be called *Brithopus*) and *Deuterosaurus,* show throughout their structure characters which can only have been derived from sphenacodont ancestors, and prove that Romer's conclusion was perfectly justified. The very important early forms still lack a detailed description, which is, however, being prepared by Professor Efremov; the drawings for the illustrations already exist.

I still feel that the Cynodontia have been derived from gorgonopsids, though Broom, partly on the evidence afforded by new material referred in part to the genus *Procynosuchus,* now tends to believe that they may have come from Therocephalia. It is becoming more and more obvious that the division of the South African carnivorous theriodonts into gorgonopsids and Therocephalia is inadequate and may be misleading; the groups as a whole lived side by side in the same country for a period presumably of the order of fifteen or twenty million years. They include creatures with a skull length of little more than an inch, and others with a skull perhaps twenty-five times that length. They range over a very wide series of adaptations, and it would not be surprising were the groups to approach in complexity that of the eutherian mammals. It is obvious that a satisfactory natural classification will require not only very many years of work but also far more material than is yet available, large though that be. At any rate I still see no reason for doubting that the story of cynodont evolution set out in Chapter V is fundamentally sound. It may be added that it has recently been shown that *Tritylodon,* long famous as the earliest mammal skull, is actually a reptile, an ictidosaurid of cynodont derivation.

VI. THE APPROACH TO A
MAMMAL STRUCTURE

IN PREVIOUS CHAPTERS I have shown that the fundamental evolution of the amphibia takes place in accordance with a series of trends, the various groups of amphibia pursuing a course of structural change which at very different rates leads them ultimately to very similar forms. During these changes the functions of each different part of the skeleton have remained the same, although the arrangement of all parts of the skull has been altered so that the whole remains mechanically sound. The changes are in proportion, not fundamentally in function. But even among the amphibia, although the carrying out to the logical conclusion of the evolutionary trends imposes certain habits on the animals in that the latest members of the majority of the stocks are necessarily aquatic, sometimes perennibranchiate, there is what Osborn called an adaptive radiation. The animals, without any change of their evolutionary grade, may adopt and become fitted for specific and sometimes very unexpected habits of life, locomotion, and food.

There are however cases in which the working out of an evolutionary trend has brought about a complete transformation of the functions of some parts of the structure; and the study of an example is useful in all consideration of evolutionary mechanisms.

In the last chapter I showed that the mammal-like reptiles, from a beginning in animals very little removed structurally from the earliest labyrinthodonts, progress in such a way that they steadily approach a mammalian structure not only in all parts of their skeleton but in the nature of their special sense organs and of their brain. They become agile, quick-moving animals with a soft, muscular, and sensitive skin, with highly developed olfactory organs, and with a well-developed series of turbinal cartilages which suggests that they were warm blooded.

As one of the most characteristic peculiarities of mammals is their dependence on hearing for warning of the approach of enemies and of food, it is interesting to investigate the nature of this sense in the mammal-like reptiles. In 1600 or 1601 (there is a dispute about the exact date of publication) Julius Casserius published one of the first works on comparative anatomy, a magnificent folio volume on the organ of voice. In it as an addendum he included a section on the organ of hearing. He showed that the goose differed from all the mammals he had investigated in that it possessed one bone in the ear drum, while the mammals had three (Fig. 54). His descriptions and

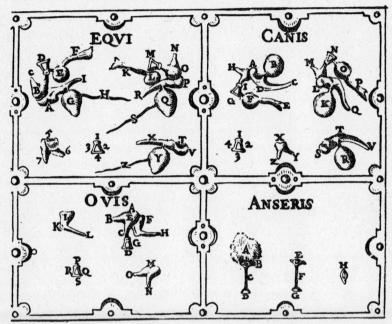

FIGURE 54. The first published drawings of the ear ossicles of mammals and a bird, from J. Casserius, 1601. In the figure labeled *Ovis,* of the sheep's ossicles, E is the incus; A its articulation with the malleus F; and C its articulation with the stapes G, whose footplate for insertion into the fenestra ovalis is D. *Anseris* is the single stapes F of the goose, with a footplate G and an attachment to the tympanic membrane E.

especially his figures are admirable; it is not easy, even now that we know the structure and have the advantages of better artificial lighting and of microscopes and lenses, to make good dissections of these parts, yet Casserius with none of these things made out an unknown structure. It is a pity that his name has come down to us in a garbled form, the Gasserian ganglion known to all students of biology being called after him.

The ear when we first meet with it in lampreys is an organ consisting of a liquid-filled sac, whose form is so elaborate that it is really divided into two parts of different function. The lower part is concerned with the determination of the animal's position with respect to a vertical axis, it is actuated by gravity and works at all times whether the animal is at rest or moving. The other part, the semicircular canals and their connections, is an apparatus for determining the direction of movement of the head, these movements, being, in all vertebrates above the cyclostomes, referred to three axes at right angles to one another. This portion is retained almost in its original form by all vertebrates, and it is probable that to our possession of it we owe our habit of regarding all objects as possessing height, width, and depth, measured along straight lines at right angles to one another.

The lower part, although primarily concerned only with position is, because of the nature of its sensory organs, readily converted into an apparatus for the recognition of sound. Professor von Frisch has shown that even some fish, goldfish for example, can appreciate and distinguish between different musical notes.

The amphibia are, however, the first group of vertebrates in which hearing became universal, presumably because they were the first colonists of the land. In air, as in water, sound is conveyed by longitudinal waves which, beating upon the outside skin, set up regular vibrations capable of activating suitable sense organs; but the energy which can be conveyed by such waves is very small, and it is obviously desirable to magnify the movements they induce. In the amphibia this is done quite simply. In them, as in all air-breathing animals, the mouth and the pharynx lying behind it are filled with air. During development this pharynx develops a series of outgrowths, some of which open to the exterior to form gill slits, while the first of the series (that which becomes the spiracle of the dogfish) always remains closed, although an outgrowth from it approaches the surface skin very closely. The thin sheet of tissue so formed, lying between the outside air and the air-filled spiracular cavity, is the tympanic membrane, which can be thrown into vibration by sound waves falling on it, and as it is light in weight can move through relatively considerable distances. It thus acts as a receptor of sound waves. It lies quite superficially, separated from the ear itself by the air-filled spiracular space, the tympanic chamber or cavity of the middle ear. Its vibrations can only be conveyed to the inner ear through some rigid rod, and this is made out of the upper element, the hyomandibular bone, of the skeleton which lies between the spiracle and the first gill slit. In the labyrinthodonts this rod, the stapes, is very heavy (about as thick as a lead pencil, and two inches long in a skull fifteen inches in length); its inner end is flat and is attached to a membrane stretched across a window, the fenestra ovalis, in the outer wall of the ear capsule. This lies immediately outside the lower part of the inner ear, which is the part concerned with hearing.

In a labyrinthodont such as *Cyclotosaurus* the tympanic membrane may have an area of about 8.5 square centimeters, and the fenestra ovalis perhaps 1.5 square centimeters. That is, the pressures exerted on the inner membrane will be five and a half times as great as those on the external tympanic membrane, if the amplitude of the movements be the same in each. But actually the tympanic membrane is horizontal, that closing the fenestra ovalis is vertical, and the stapes is a bell-crank lever with one limb only a fifth as long as the other. Thus the actual excursion at the fenestra is only one-fifth that of the drum and the resulting pressures about twenty-five times as large.

As I showed in the last chapter the mammal-like reptiles lost this apparently excellent arrangement. In them the end of the stapes becomes firmly articulated with the quadrate bone, and the tympanic membrane is driven from its original position and in some cases appears to have vanished altogether. But in mammals there is a tympanic membrane which is connected with the fenestra ovalis, not by a single stapes but by an articulated chain of three or, more accurately, four bones. The problem before us is an old one; it is to explain the significance of this difference and to discover how it arose.

If we compare the skull and lower jaw of any recent reptile with that of a mammal (Fig. 55), we find that one of the most obtrusive differences is that whereas the reptile lower jaw is built up from several bones that of a mammal is single. The reptile jaw articulates with a special bone (the quadrate) in the skull, while in the mammal it meets the squamosal in an entirely different way, articulating with it by a rounded condyle.

In the lizard (Figs. 55 and 56) the tympanic membrane lies on the side of the head, depressed below the surface for its protection so that it forms the floor of a shallow pit, but easily visible from the outside. Its anterior border is stretched on a bony frame made up of the quadrate bone above and that part of the lower jaw which projects behind the articulation for its lower part. The hinder edge of the membrane is supported by a muscle which passes down from the upper part of the skull to the tip of the retroarticular part of the jaw. It is thus evident that the tension in the tympanic membrane

MAMMAL

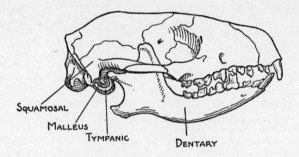

SQUAMOSAL

MALLEUS

TYMPANIC   DENTARY

LIZARD

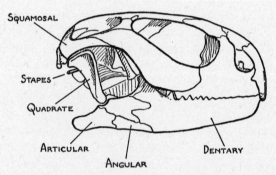

SQUAMOSAL

STAPES

QUADRATE

ARTICULAR   DENTARY

ANGULAR

FIGURE 55. The skull and lower jaw of a mammal, the
hedgehog, compared with that of a lizard. To show that
the mammalian lower jaw is a single bone, the dentary,
articulating directly with the squamosal bone; while that
of the reptile is complex, including an angular, and an
articular bone articulating with a quadrate which itself
is attached to the squamosal. In the mammal the tympanic
membrane is stretched across a tympanic bone and a
malleus is attached to it. In the lizard the tympanic membrane is supported by the quadrate and the articular and
is directly attached to the end of the stapes through a
cartilaginous extracolumella.

will be altered when the mouth is opened or closed, a condition
which must interfere with the sense of hearing. The vibrations into
which the tympanic membrane is thrown by sound are conveyed
to the fenestra ovalis by a stapes or columella, whose inner end,
sometimes perforated by a foramen and expanded into a footplate

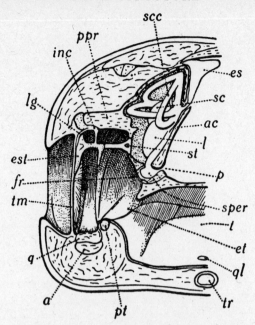

FIGURE 56. Left half of the head of a lizard cut transversely through the tympanic cavity to show *st.*, columella; *inc.*, *lg.*, and *est.*, which are processes of the extracolumella; and *tm.*, which is the tympanic membrane. From Goodrich.

which fills the fenestra ovalis, is bony, while the rather elaborate outer end, the extracolumella, remains cartilaginous. This extracolumella is attached to the tympanic membrane excentrically, and from it there arises a dorsal process which is attached to the extremity of the paroccipital process, the lateral extension of the ear capsule. A similar lateral process from the columella passes downward, to become the main part of the hyoid-arch skeleton in some reptiles or to be attached to the quadrate in many lizards.

In mammals the conditions are quite different (Fig. 57). In them the tympanic membrane has sunk deeply into the head, behind and below the articulation of the jaw, and lies at the inner end of a long tube (the external auditory meatus) whose outer opening is usually embraced by the base of the pinna or external ear. This is a trumpet designed to concentrate sound waves by reflection so that less intense noises may produce vibrations in the membrane sufficiently powerful to activate the sense organs of the organ of hearing. The pinna is usually movable, there being special muscles for that purpose derived from the skin muscles, which are a mammalian peculiarity, and from the facial musculature, which also only occurs in them. The movable pinna is important because, taken with the separation in space of the two external openings of the outer ear and their symmetrical position, it enables hearing to be used in determining the direction from which sounds are coming, an ability of very great value to mammals.

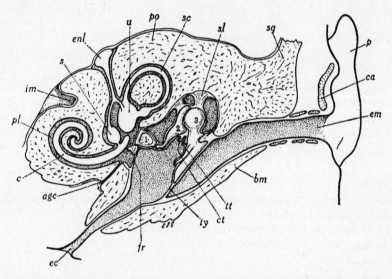

FIGURE 57. A diagrammatic representation of the ear of a mammal, to show the three ear ossicles: 1) the stapes, 2) the incus, 3) the malleus with its insertion into the tympanic membrane, *ty*. From Goodrich.

The tympanic membrane is supported entirely by a special horseshoe-shaped bone, the tympanic, whose opening may be very small and lies posteriorly. In higher mammals the tympanic bone becomes involved with others to form a bony shell, the tympanic bulla, which maintains the shape of the air-filled middle ear or tympanic cavity. The tympanic membrane is attached to one leg, the manubrium, of a small two-legged bone, the malleus. The other extremity stretches forward and inward within the tympanic cavity. A small but very important muscle, the musculus tensor tympani, is attached to the base of this, the processus folianus, and a nerve, the chorda tympani, runs forward along its inner surface, and may in some mammals actually penetrate it.

The meeting point of the two limbs of the malleus is an expanded and usually rounded knob, which is strung up by a ligament to the roof of the tympanic cavity and bears on its inner surface a face whereby it is firmly but movably articulated to another bone, the incus. This bone is in some mammals supported by actual contact with the roof of the tympanic cavity through a short process which is always a fixed point. From its body a slender arm projects downward whose inner surface is movably connected with the end of the stapes, usually through a separate little bone, the os orbiculare.

The stapes is a short, usually stirrup-shaped bone whose footplate is inserted into the membrane closing the fenestra ovalis. The whole apparatus is designed so that vibrations of the tympanic membrane, which usually faces downward and outward, are transmitted to the footplate of the stapes, that bone rocking to and fro on the end of the long axis of its oval footplate. The amplitude of these vibrations is much reduced during their transmission and the footplate is of very much smaller area than the tympanic membrane, so that the pressures exerted are magnified.

The whole apparatus is vastly smaller and lighter even in large mammals than in the labyrinthodonts, or indeed in primitive reptiles, and it is thus probable that sounds of very much less intensity are audible.

If the mammal-like reptiles are the ancestors of the mammals it is obvious that in the great series known from South Africa we should find all stages in the transformation of the reptilian apparatus to that of mammals. *Seymouria,* which reproduces the structure of the most primitive reptiles, has a skull essentially like that of a primitive embolomerous labyrinthodont. In it the quadrate, with which the lower jaw articulates, is a large bone held firmly between two membrane bones (the quadratojugal and squamosal) on the outer surface of the head and another (the pterygoid) on its inner surface. The cartilage in which the quadrate was formed extends forward for a very long distance and, immediately above the basal articulation with the brain case, contains an epipterygoid bone (Fig. 58). There is a long otic notch across which the tympanic membrane was stretched, and the end of the stapes was attached to the membrane as it is in all later reptiles and birds except the members of the mammal-like reptilian line.

In *Captorhinus,* a primitive animal on the mammal side of the division (Fig. 58), the quadrate is still very large and the epipterygoid as it was in *Seymouria,* but that obliteration of the otic notch which is diagnostic of the mammal-like reptiles has brought the quadrate into contact with the lateral end of the paroccipital process of the ear capsule, so that it is adequately supported by this contact and by the squamosal and quadratojugal, the pterygoid having become much less important mechanically than it formerly was. In *Captorhinus* the downward rotation of the paroccipital process has driven the stapes before it until it has come to articulate firmly with the inner surface of the quadrate. The shapes of all the structures concerned make it improbable that *Captorhinus* had a tympanic membrane; it must have heard by transmission through the bones of the head.

The first of the pelycosaurs, the Lower Permian mammal-like reptiles proper, differ little from *Captorhinus* in their skulls. Only the appearance of a temporal vacuity, which enables the masticatory muscles to thicken during their shortening without bringing about serious movements of other structures contained within the skull, shows their systematic position. In the more primitive forms like *Varanosaurus* the lower jaw is considerably compressed from side to side and the hinder part, made up of the surangular and angular bones, becomes very deep.

In more advanced pelycosaurs, such as *Dimetrodon* (Fig. 59), lateral compression and deepening are carried still further. This has the result of deepening also the hinder part of the tooth-bearing bone on the outer surface of the jaw, the dentary, whose posterior extremity is directed upward and lies in a groove just below the sum-

FIGURES 58, 59, and 60. Drawings of the outer side of the lower jaws, and of the pterygoid, epipterygoid, and quadrate, of a series of skulls, usually shown with the mouth open. In all the figures the pterygoid is covered by vertical lines, the quadrate dotted, and the angular has an oblique hatching. The series from *Dimetrodon* onward is placed in order of time and shows the steady reduction in size of the quadrate and the back of the lower jaw. Concurrently the dentary expands, rising to a coronoid process above the hinder part of the jaw, and in *Cynognathus* and *Protacmon* spreading backward over the summit of the hinder part of the jaw toward, but laterally of, the quadrate. In addition the hinder part of the pterygoid is reduced, first in depth and then in length, until it is replaced by a process from the epipterygoid, which itself is reduced in *Protacmon*. The reflected lamina of the angular becomes larger up to *Arctognathus* and is then reduced in width, so that in *Protacmon* it is directed downward at an angle to the body of the bone, and is evidently homologous with the bone in *Didelphys* which in the adult becomes the tympanic. The last figure, whose resemblance to *Protacmon* is very striking, is of a pouch young of the opossum *Didelphys*.

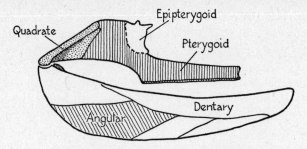

EMBOLOMERID
U.Carb.

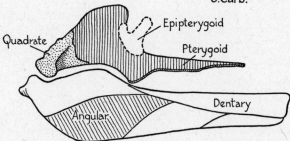

SEYMOURIA
L.Perm.

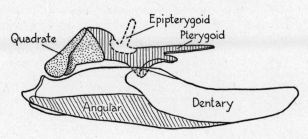

CAPTORHINUS
L.Perm.

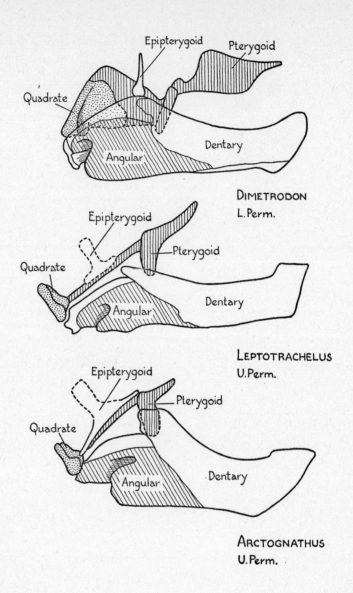

DIMETRODON
L. Perm.

LEPTOTRACHELUS
U. Perm.

ARCTOGNATHUS
U. Perm.

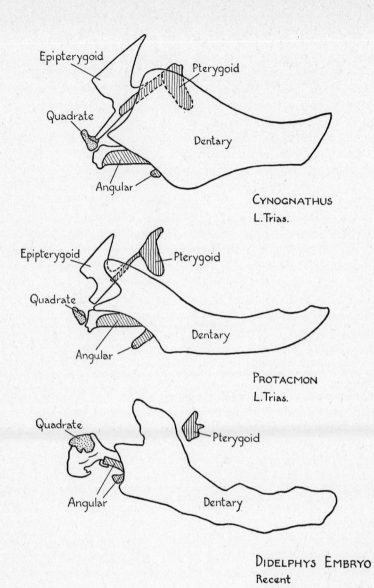

CYNOGNATHUS
L. Trias.

PROTACMON
L. Trias.

DIDELPHYS EMBRYO
Recent

mit of the surangular. At this stage, so far as our present knowledge goes, there appears a new character. The lower border of the angular bone is notched so that a special small lappet, the reflected lamina, is isolated and moves outward to cover the outer side of a small, closed pocket which opens backward and downward below the main body of the articular region of the jaw. I can see no reason for the existence of the reflected lamina except as a support for the outer surface of a cavity, and a cavity in this position can only be a diverticulum of the middle ear produced by downward and forward growth from the hyoid pouch. Such a cavity would be comparable with the air-filled spaces which are supposed to act as resonators in many mammalian (e.g., Typotherium) skulls. But the reflected lamina has a smooth upper margin, clear of all other structures, and it is conceivable that it supported a small tympanic membrane, stretched over the hinder part of this cavity and extending backward to a musculus depressor mandibuli, as does that of a lizard. If so we must assume the presence in the live animal of a small cartilaginous extracolumella continuing the stapes to the hinder part of the membrane. The structure of this region of the skull is almost always deformed by crushing in known fossils, and I am unable to make up my mind as to whether such an arrangement could actually have occurred.

The mammal-like reptiles immediately succeeding the pelycosaurs are known only from the inadequate descriptions of fossils from the copper-bearing sandstones of the Ural Mountains, but these give enough information to enable us to understand the way in which the most primitive South African mammal-like reptiles arose from a primitive pelycosaur stock.

If we consider the skull of an Upper Permian gorgonopsid (Figs. 61–65) we find that the whole head is much shallower than in pelycosaurs, and that this decrease is largely due to a thinning of the floor of the brain case and a decrease of all the structures which lie below the level of the base of the brain. These animals were carnivorous, and their dentition had become sharply differentiated into incisors, large canines, and cheek teeth. So powerful a dentition obviously requires corresponding musculature.

All pelycosaurs have flat-sided skulls. The cheek is vertical and the space for temporal muscles is constricted not only by the zygomatic arch which lies below the single temporal vacuity but also by the extension of the pterygoid backward to the quadrate and by the wide separation of the articular facets for these bones which are

FIGURES 61–65. Drawings of the skull and the outer side of the lower jaw of a gorgonopsid reptile near to *Scylacops*. To show the general structure of an advanced mammal-like reptile, and especially the bowing out of the zygomatic arch (squamosal in Figs. 61, 62 and 64), and the reduction of the quadrate to a small bone placed in a recess on the front face of the root of the squamosal. The stapes, and its continuation outward behind the quadrate, is shown in Figs. 63 and 64. In Fig. 61 the concave hinder surface of the quadrate, adapted to the tympanic membrane, is well shown. In Fig. 65 the reference line from the legend "Angular" ends on the upper limit of the reflected lamina. Original.

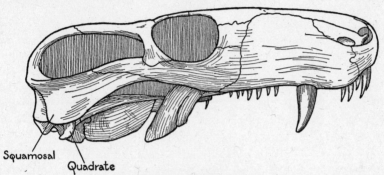

FIGURE 61. *Scylacops*, right-lateral aspect

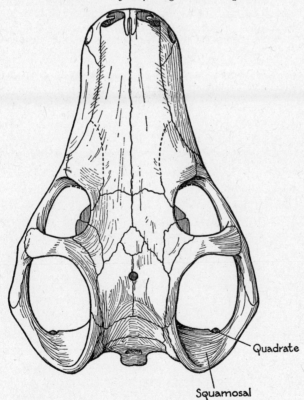

FIGURE 62. *Scylacops*, dorsal view

149

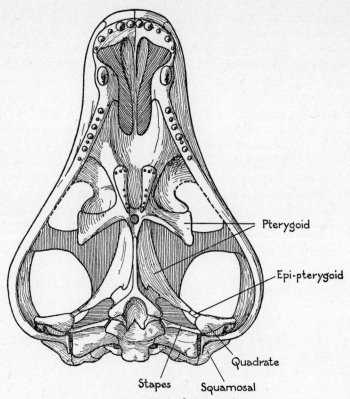

FIGURE 63. *Scylacops*, palate

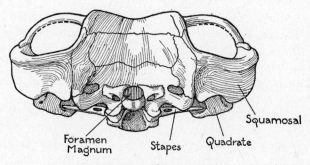

FIGURE 64. *Scylacops*, occiput

150

attached to the base of the neural cranium. In pelycosaurs the temporal muscles lay entirely on the inner side of the hinder part of the lower jaw, only a few of the outer fibers of the muscle being attached to the summit of the dentary bone, and none passing down onto the outer surface of the jaw.

These relationships of muscle and bone require a deep skull in order that the range of action of the muscle may be great enough to enable the mouth to take in large prey. In the gorgonopsians, which first appear in bottom Upper Permian age rocks at the base of the fossiliferous succession in South Africa, the transverse sectional area of the muscles (which determines the pull they can exert) is increased by a bowing outward of the zygomatic arch and by a migration downward of the muscle attachments on the outer side of the lower jaw which ultimately leads to the isolation of one part of

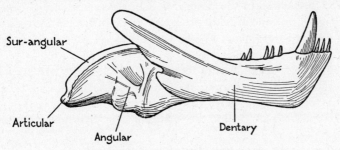

FIGURE 65. *Scylacops,* outer side right lower jaw

the originally single muscle mass as an independent masseter muscle.

At the same time the space at the disposal of the inner part of the biting muscles—the temporal muscles, which come down from the skull roof, and the pterygoidal muscles, which pass backward and downward from the palate—is enlarged at the expense of a reduction in size of the ramus of the pterygoid which passes backward to the quadrate, and by the forcing of this bone inward until it is plastered onto the base of the brain case in such a way that the whole forms a complex, narrow, but extremely stiff girder.

Concurrently the attachment area of the origin of the great temporal muscle is increased by the backward extension of the hinder border of the temporal vacuity, which is formed by the squamosal bone. The fibers of this part of the muscle lie at only a small angle to the horizontal and are short. Their range of action is a percentage of their total length, and they can lengthen sufficiently to allow the mouth to open to the desired extent only if they are attached to the short arm of a lever and act at a mechanical disadvantage. The necessary modification of the lower jaw is brought about by the growth upward and backward of the extreme posterior tip of the dentary bone so that it rises freely above the upper margin of the surangular.

Concurrently with these changes the quadrate becomes relatively smaller. It has no mechanical function other than the transmission of pressures, exerted on it by the lower jaw, to the squamosal bone and thus to the brain case. Therefore the quadrate retains its original width, and with a much reduced quadratojugal is left behind when the zygomatic arch bows outward. The two small bones fuse completely. The upper end of the quadrate complex is a thick, rounded margin fitting accurately into a depression on the front face of the root of the squamosal, just lateral to the articulation of that bone with the end of the paroccipital process. A thinning flange of squamosal grows steadily down the posterior surface of the quadrate, but even in the original of Fig. 64 (an advanced gorgonopsid) leaves nearly half of it exposed. The quadrate is held in position by a backward extension of the base of the epipterygoid, which reaches its anterior face at the inner corner; and is itself supported and held in position by the posterior ramus of the pterygoid, which now no longer extends back to pass behind the quadrate as it does in all more primitive reptiles.

The relatively small posterior part of the lower jaw, whose outer surface is composed of the surangular and angular bones, still exhibits the notch and reflected lamina which appeared de novo in the advanced pelycosaurs, but the reflected lamina has grown upward so that it hides a good deal of the outer surface of the jaw.

The stapes passes across from the fenestra ovalis to the quadrate and is firmly tied to that bone, this being the typical mammal-like reptile condition. But in the original of my figures (a gorgonopsid) there is a slender bony extension of the stapes, which passes outward free from contact with other bones, and is directed toward the center of curvature of a deep cylindrical hollow horizontally crossing the posterior surface of the quadrate. It seems evident that there was a small tympanic membrane attached to the quadrate and closing this groove. That this was indeed the case is confirmed by the ridge which bounds the occiput and ends below in a border suitably placed to serve as an attachment for the upper edge of the membrane, and by the existence of a groove bounded on its inner side by this ridge and extending upward and outward to the side of the head. This groove seems to have housed a very wide and short external auditory meatus.

If this conclusion be justified, as seems probable, we know the general extent of the tympanic cavity. It extended inward and forward below the paroccipital process, opening to the pharynx by a Eustachian tube which passed downward over the notch where the quadrate ramus of the pterygoid left the floor of the brain case, a point marking the original place of articulation between the two structures.

Nevertheless the shape of all the bones concerned makes it perfectly possible for there to have been an anterior and ventral prolongation of the tympanic cavity which would end in the pocket between the reflected lamina of the angular and the rest of the lower jaw. The upper part of this diverticulum would lie in a groove on the ventral surface of the hinder part of the jaw, and its hinder end would pass laterally to a special descending process of the articular bone to open into the main cavity immediately mesial to the tympanic membrane. The descending process of the articular is highly variable in size (Fig. 66) but it appears to have served for the attachment of a muscle which pulled the jaw downward, thus opening the mouth.

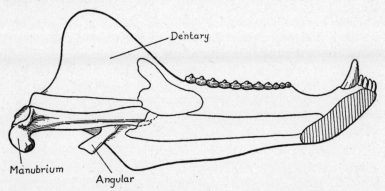

FIGURE 66. The inner surface of the lower jaw of *Gomphognathus*. To show the extent of the dentary and the character of the reflected lamina of the angular, on which the reference line ends. The manubrium of the articular bone is important. Original from specimens from Lady Frere, Cape Province, South Africa.

One other bone demands a short description. The epipterygoid in Gorgonopsians is usually made up of a narrow but flattened rod whose upper part reaches the skull roof while ventrally it rests upon the upper surface and the lateral extremity of the basipterygoid process of the floor of the brain case, which is the primitive articulation with the pterygoid. From the base of the bone a slender process passes backward above the quadrate ramus of the pterygoid toward the quadrate.

The fate of all these structures can be followed in South African specimens. In them there is a continuous history from the early gorgonopsids to the late cynodonts, but unfortunately much of the material is still inadequately prepared and described and it is difficult at present to isolate the many existing lines of descent. Thus the series of animals illustrated in Figs. 58–60 is artificial in that, although they are arranged in correct morphological order and are never out of place in time, they are not parents and children, and in some cases (especially in the earlier forms) are actually members of side branches which retain structural conditions long abandoned by the contemporary representatives of the main stock.

If, however, we begin with the Upper Permian gorgonopsid *Leptotrachelus* we can follow the story easily, and to all appearance truly. *Leptotrachelus* is earlier in date than the unnamed gorgonopsid described above and is correspondingly less advanced. The quadrate complex is relatively large and is still reached by the pterygoid, though the quadrate ramus of the epipterygoid does not extend so far as to come into contact with it. The notch in the angular lies far back and the reflected lamina is short and small. The dentary has only a very small free point above the surangular.

In *Arctognathus,* which is later in time than the unnamed gorgonopsid, the quadrate is small and the pterygoid does not reach it, although it is probable that the epipterygoid did. The notch in the angular has moved forward, and the reflected lamina is a large sheet of bone extending down below the lower border of the rest of the hinder part of the jaw. The free upstanding coronoid process of the dentary has become much larger, but the hinder border of that bone still passes downward and forward nearly in a straight line to the lower border of the jaw.

In the Lower Triassic cynodont, *Cynognathus,* all the changes are in the same direction but carried much further. The quadrate is much smaller and the quadrate ramus of the pterygoid has completely vanished, while that of the epipterygoid is still complete, resting against the front face of the quadrate. The notch in the angular has moved so far forward that its anterior end is covered by the dentary, so that the short reflected lamina projects downward and backward entirely below the lower border of the rest of the hinder part of the jaw. The dentary has expanded enormously. Not only is the coronoid process both high and wide but the posterior border of the bone below it has grown backward over the hinder part of the jaw, now very slender, so that its posterior point, which lies immediately dorsal to the surangular, fails to touch the quadrate by a very short distance.

In *Protacmon,* which although more advanced is a contemporary of *Cynognathus,* the quadrate complex is extremely small and the quadrate ramus of the epipterygoid has undergone a reduction exactly paralleling that of the quadrate ramus of the pterygoid. It no longer meets the quadrate, indeed it does not approach it at all nearly. The hinder part of the lower jaw is now an extremely slender structure. The V-shaped notch in the angular, whose lower border is formed by the straight, slender reflected lamina, is wide and very far forward. The dentary is still further developed. The high but slender coronoid process passes far back into the temporal fossa; below it the margin of the bone is cut out into a bay bounded by a point which extends backward, in contact with the upper surface of the surangular, until it fails to touch the quadrate by only some two or three millimeters. This point is still quite thin and evidently played no part in the articulation between the jaw and skull.

For our next stage we must go to the head of a pouch young of any marsupial, *Didelphys* (the opossum) or *Perameles* (the bandicoot) being the best known. Direct comparison between the drawings of *Protacmon* and *Didelphys* (Fig. 60) will show that the dentary bones are very similar, not only in outline but in form, and that they have identical relations to the hinder parts of the lower jaws. The quadrate of the marsupial is different in shape from that of the reptile but is of very much the same proportionate size. In both it articulates with the outer end of a rodlike stapes, and in both the greater part of its anterior surface is a facet whereby it articulated with the hinder end of the articular of the lower jaw. In both animals the quadrate is in contact with the squamosal alone, the epipterygoid (and its homologue the alisphenoid) forming the front wall of the tympanic cavity. In reptile and marsupial alike the angular bone is split into two limbs, the upper forming part of the functional posterior portion of the lower jaw and the lower being a descending reflected lamina.

But the structures in *Didelphys* which I have called quadrate, articular, and angular because they are obviously the homologues of these bones in the cynodonts become the ear ossicles, the incus, malleus, and tympanic, respectively, of the adult marsupial. Indeed the marsupial is born with a reptilian jaw and jaw suspension, it actually opens and closes its mouth on the old primitive articulation between quadrate and articular, i.e., between incus and malleus, and only later, when a backward growth of the dentary brings that bone into contact with the squamosal, are the ear ossicles freed from their historic function and devoted exclusively to the sense of hearing.

One problem remains for consideration. We must discuss whether the mammalian tympanic membrane is or is not the same as that very small structure which we saw reason to believe existed in gorgonopsids, lying behind the quadrate and connected to the main body of the stapes by a slender extracolumella. That this extracolumella was of general occurrence seems to be established, because Dr. Broom has described it in a therocephalian only rather remotely related to the Gorgonopsians, and Mr. Parrington has found it in a primitive cynodont.

If we compare Figs. 66 and 67, representing the inner surface of the jaw of the cynodont *Gomphognathus* and of a pouch young of the marsupial *Perameles,* we can see the great resemblances recorded above. In addition there is to be found in the cynodont a powerful ventral process from the articular bone which appears to have been for the insertion of a muscle for opening the mouth and is of inconstant appearance in members of the group. This process is very similar in its position and general appearance to that part of the manubrium of the mammalian malleus which is inserted into the tympanic membrane.

If now we consider the skull and lower jaw of a cynodont as a whole (Fig. 68) on the basis of a comparison with gorgonopsids and mammals, we shall find it possible to determine the actual position of most of the soft parts of the head.

Inspection of actual specimens shows conclusively that by the time evolutionary change had reached the point represented by *Cynognathus* all the important masticatory muscles were attached to the dentary bone. The masseter, clearly an independent muscle arising from the zygomatic arch and extending forward below the eyes, must have had an insertion on the dentary toward its lower border. It must have pulled nearly vertically, so that only a small horizontal component existed which forced the articular backward against the quadrate. The great temporal muscle inserted on the summit of the coronoid process and on both surfaces of the hinder part

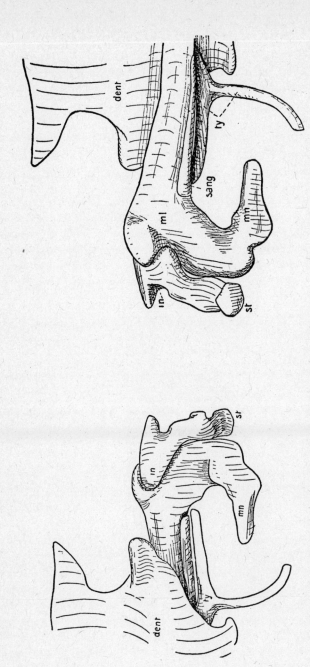

FIGURE 67. The inner and outer surfaces of the hinder part of the lower jaw and ear ossicles of a pouch young of the marsupial mammal *Perameles*. This should be compared with Fig. 66. *Ty.*, the tympanic, is the homologue of the angular; *mn.* is the manubrium; *in.* is the incus, homologous with the quadrate; *st.* the stapes. From Palmer.

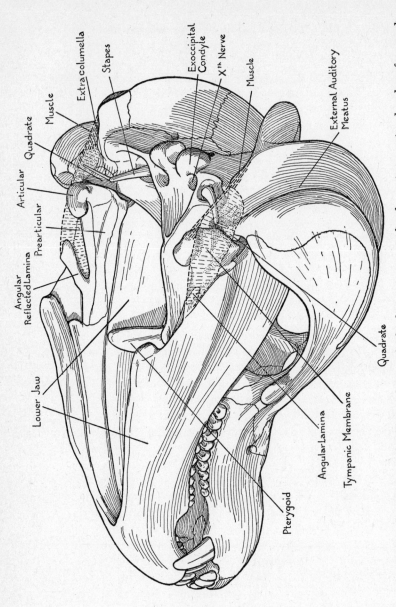

Angular Reflected Lamina

Articular

Prearticular

Quadrate

Muscle

Extracolumella

Stapes

Exoccipital Condyle

Xth Nerve

Muscle

External Auditory Meatus

Lower Jaw

Pterygoid

Angular Lamina

Tympanic Membrane

Quadrate

FIGURE 68. The skull of *Gomphognathus*, with the lower jaw articulated, resting on its dorsal surface and viewed obliquely from below and behind. To show the position of the stapes and the hypothetical extracolumella. The external auditory meatus fixes the position of the tympanic membrane, and the drawing shows that this may have extended forward laterally to a diverticulum of the tympanic cavity, so that it is supported by the reflected lamina of the angular and the manubrial process of the articular, its hinder border being carried by a musculus depressor mandibuli.

of the dentary has, I think, inevitably a strong horizontal component in its pull. The pterygoid muscles originally attached to the angular on the lower border of the hinder part of the jaw seem largely to have migrated onto the posterior point of the ventral border of the dentary, which is thickened and roughened as a muscle insertion, but there is no doubt that part of them still retained the old insertion, becoming in mammals the musculus tensor tympani, attached to the malleus.

The horizontal component of the pull of this pterygoid muscle seems to have been directed forward. It opposes that of the temporal muscle, and must have reduced the force tending to push the articular against the quadrate bone and slide the whole very reduced hinder part of the jaw forward along the groove on the inner surface of the dentary in which it rests.

Thus the hinder part of the jaw is freed from most muscle insertions. In cynodonts (see Figs. 53 and 68) the outer surface of the squamosal is impressed by a groove which, beginning far forward on the summit of the skull, passes backward and downward until it ends at a ridge, just lateral to the end of the paroccipital process, in the middle of the width of the quadrate complex, and at the point where the powerful crest which bounds the occiput abruptly ends. This groove in some aged specimens of *Gomphognathus* is so greatly overhung by its interior wall that it is almost converted into a tube, and it is certain that it housed the external auditory meatus. The existence and position of the tympanic membrane are thus certain.

If we accept the view that the reflected lamina of the angular was associated with a diverticulum of the tympanic cavity, we can go on to consider where this space lay in cynodonts. Inspection of actual specimens, or of Fig. 68, shows that it must have passed forward over a horizontal groove, which crosses the outer surface of the muscular process of the articular bone, arising from the tympanic cavity immediately within the membrane. If this were the case, then the thin outer wall of the diverticulum would lie on the same plane as, and be directly continuous with, the old tympanic membrane. Its borders would be supported by the two limbs of the angular and by the outer surface of the muscular process of the articular. When by further backward growth the dentary came to meet the squamosal at a point immediately lateral to the articulation of the quadrate and articular, the two parts of the jaw, though independently supported, moved together without any displacement of one on the other.

But the direct contact of dentary and squamosal is so obviously superior mechanically that it seems reasonable to assume that all the hinder part of the jaw would very soon be reduced in size by failure to grow after an early age, and would thus be available for complete subservience to the sense of hearing, with whose apparatus these bones had long been involved.

The character of the membranous labyrinth, the actual organ of hearing, in cynodonts can be judged from the shape of the cavity within which it lay. It is evident that there was in these reptiles a cochlea, directed forward and inward in a mammalian manner, and bent through a quadrant as is that of the most primitive recent mammal, the monotreme *Ornithorhynchus*. The cynodont sense of hearing must thus have been essentially mammalian, but the excessive weight of the stapes, which is a rod about one inch long and an eighth of an inch in diameter, and the fact that the disposition of the parts renders it likely that movements of the small tympanic membrane were transmitted without reduction of amplitude to a fenestra ovalis of large size suggest that only sounds of high intensity can have actuated the receptor organ. With the increase in area of the tympanic membrane which resulted from the conversion of the outer wall of the anterior diverticulum of the cavity, and the possibility of reduction in amplitude of the movements of the membrane closing the fenestra ovalis, which was presented by the existence of a chain of bones (the articular, quadrate, and stapes) connecting the two membranes, it seems justifiable to assume that the activity of hearing became greatly increased and that that sense came to play a much larger part in the life of the animal.

The story of the origin of mammals thus seems to differ very greatly from that of the origin of reptiles. In the latter we found in *Seymouria* an animal which, in the fundamental matter of its life history, appears to have been an amphibian, still tied to water by the preservation of a larval period spent necessarily in pools or rivers. But nonetheless the group of animals of which *Seymouria* is a member is distinguished from all true amphibia because during the long period from Upper Carboniferous to Upper Permian time it shows no trace of any of the evolutionary changes which follow the trends common to all orders of amphibia. *Seymouria* possesses a small series of characters, none of any discoverable significance, which are to be found only in reptiles. Although too specialized in some ways, and much too late in time to have been ancestral to reptiles as a whole, it nonetheless does preserve for us most of the qualities of the most primitive reptiles and allows us to speculate on the mode of origin of that group.

It seems evident that the reptiles arose "suddenly" from a very early stage of the amphibian order Labyrinthodonta and that they were at first distinguished from their contemporaries only by their possession of trivial characters which seem to have accumulated one by one as a result of the chance association in a single animal of a group of unrelated genes.

Mammals on the contrary seem to have arisen gradually by the steady, progressive modification of all the parts of their bodies. These changes in many cases, the limbs, for example, proceed in such a way that the mechanical efficiency of the animal is increased.

It is indeed probable that many of the more fundamental peculiarities of mammals—their hair-covered, scaleless, and muscular skin, and the stable body temperature, which enables them to disregard external climates, for example—may have been acquired while in all technical points of skeletal structure they were still reptiles.

The evidence I have adduced in this chapter shows that the characteristic structure of the mammalian middle ear arose by the conversion to an entirely different function of bones which, having been superseded by a new device, would normally have disappeared from later members of the group were it not that their spacial position and involvement with an extension of the tympanic cavity enabled them to be converted to the entirely new function of a subservience to the sense of hearing.

APPENDIX

Since the sixth lecture was delivered Westoll, in two small notes (1944 and 1945) has quite independently reached conclusions which agree in essence with those I set out in 1937. Romer has shown that the notch and reflected lamina of the angular bone of pelycosaurs is restricted to and first appeared in the sphenacodonts, of which *Dimetrodon* is the most advanced member, and exists in them although it now seems quite evident that they possessed no tympanic membrane whatsoever and that the stapes presumably lacked an extracolumella of modern reptile pattern, which seems to have been represented in labyrinthodonts in general. Thus it is very unlikely that the upper border of the reflected lamina of *Dimetrodon* did serve as an attachment to the lower border of a tympanic membrane. In a recent paper * I have accepted an early view of Romer's that the reflected lamina owes its position to a migration of the hinder end of the anterior pterygoid muscle onto the outer face of the angular, the reflected lamina so cut out surviving not because it formed the outer wall of an air-filled diverticulum of the middle ear but because it was of importance as an insertion of the intermandibular muscle.

The pocket between the reflected lamina and the body of the articular is very varied in shape and character in different mammal-like reptiles. In *Endothiodon* and its allies it may become very capacious, the reflected lamina flaring out suddenly from the rest of the bone, obviously to house some structure which, in this case, can scarcely have been an air-filled sac. In dicynodonts in general it is a narrow slit, which may well have been completely filled by the anterior pterygoideus muscle or its ligament of attachment. In some gorgonopsids and therocephalia at the point where the upper border of the reflected lamina joins the body of the bone there is a rounded border which in early figures of the skulls belonging to this group is represented as a foramen, and it is conceivable that, as Broom suggested, the parotid gland had migrated into the pocket, the notch allowing its duct to pass forward.

The immediate successors of the pelycosaurs in the copper-bearing sandstones of the Urals still await description by Professor Efremov, but he has shown me his materials and explained his interpretation of them. It is evident that the tapinocephaloid *Deuterosaurus* has a reflected lamina which arises far forward from the body of the bone and bounds only a very narrow pocket, the condition being very much as in gorgonopsids. I think it is most improbable that this pocket could, in these animals, have been occupied by an extension of the middle ear. It seems in fact certain that neither in pelycosaurs nor in these later reptiles can any extracolumella portion at all comparable with that of lizards have existed.

In the above-mentioned paper I have shown that it is possible to restore the head musculature of *Dimetrodon* with some confidence because the muscles concerned have, in many cases, left definite evidence of their attachments or position in the form of surface sculpture on the skull and jaw bones or in grooves on such bones. The most interesting feature of this musculature is the attachment of the hinder end of the anterior pterygoid muscle to the outer surface of the articular region, and the way in which that muscle winds around the lower border of the jaw to pass forward in a special, sharply defined groove. There is also a posterior pterygoid muscle

* Bibliography, 131. Watson, 1948.

attached to a special ventral process of the articular bone, quite distinct from the retroarticular process. This muscle passes inward and then suddenly upward toward the root of the epipterygoid.

In all later mammal-like reptiles the retroarticular process and the musculus depressor mandibuli which is attached to it disappear completely, while the process for attachment of the posterior pterygoid muscle, which has commonly (though incorrectly) been regarded as a retroarticular process, remains, varying greatly in size in different forms. This condition seems to give additional confirmation to the view that the tympanic membrane had completely disappeared, and this implies that it is irrational to look for relics of the lizard extracolumella in the development of the mammalian ear. The account in the lecture of the development of the temporal and masseter muscles, and of the quadrate and all the bones in relation to it, still seems to me to be correct. I have, however, made a further preparation of the gorgonopsid figured in the paper, and find that the columella ends in a rounded surface lying close down on the upper surface of the backwardly turned lower part of the quadrate condyle. It is quite evident from the material that there is no direct end-to-end contact of the columella on the quadrate, and no indication whatsoever that an extracolumella passed backward to be inserted into a tympanic membrane. Indeed, although there is a rounded notch on the posterior side of the quadrate, as shown in the figures, and although this notch, which is completed above by the lower border of that flange for the squamosal which passes backward in contact with the end of the paroccipital, could conceivably have supported a tympanic membrane there is no evidence that it did in fact do so. And indeed it seems difficult to believe that such a tympanic membrane, with a surface area of only some 10–15 sq. mm., could have been of any value when it is remembered that the end of the stapes is some 3 mm. wide and 1 mm. thick, and that the bone is tightly inserted into the cartilaginous lips of the fenestra ovalis as a plug some 4 mm. in diameter, the contact being more than 3 mm. in length. It is in fact obvious that the effective conduction of sound waves is through bone.

There is in this, as in most other gorgonopsids, a posteriorly directed flange from the squamosal which lies against the truncated outer end of the paroccipital process and, dorsally, against the reflected hinder margin of the tabular. It has usually been assumed that this flange corresponds with that forming the inner margin of what is, by common consent, a groove within which lay the external auditory meatus in the cynodont *Diademodon.* I am not however entirely convinced that the two flanges are in fact homologous; and there is, in any case, no doubt that the actual flange (as it exists in gorgonopsids) is homologous with one in the same position which is even to be found in a rudimentary form in the very archaic deinocephalian *Deuterosaurus,* and which there, as in all the other forms in which it occurs, marks the lateral limit of the area of insertion of the neck muscles. Thus the original discussion will need some modification.

The general account of the development of the epipterygoid, and all bones which come into relation with it, still seems to me essentially correct. But Broom has recently described some very remarkable forms from the cisticephalus zone which he believes to be ancestral to the cynodonts, themselves of therocephalian derivation. These animals possess a secondary palate, sometimes, according to Dr. Broom's account, not completely closed; and in the case of *Levachia* much is known of the postcranial skeleton, where the character of the pelvis is extraordinarily similar to that of a cynodont and the shoulder girdle provides a form from which the cynodont condition could readily be derived. I am not, however, clear that these procynosuchids are of therocephalian origin. As I have pointed out above the Theriodonta as a whole are a group including very many diverse stocks, and the terms Gorgonopsia and Therocephalia relate merely to the existence in association of individual primitive skull characters retained for a longer time in the gorgonopsids than in the therocephalian stocks.

I now doubt whether a tympanic membrane existed in the Gorgonopsia, although its presence in cynodonts seems to be established by the existence of the external auditory meatus. It is thus, perhaps, conceivable that such a membrane was a neomorph attached to the angular and hinder part of the lower jaw alone. The ventral process of the articular bone in cynodonts served for the attachment of the posterior pterygoid muscle, which becomes the musculus tensor tympani in mammals, and it is now evident that the mouth of cynodonts was opened by a musculature perhaps of the type found in monotremes or by a device analagous to the digastric of mammals.

VII. ADAPTIVE CHANGE

IN NONE of the cases we have so far considered have we discussed the evolution of an animal which, over a prolonged period of time, shows structural changes that can reasonably be interpreted as the result of an adaptation to some special characteristic mode of life. Such adaptation may presumably arise in two ways. It may represent the exploitation of a new opportunity, the availability, for instance, of a new kind of food leading relatively rapidly to the appearance of structures and habits which fit an animal well for life under new, changed conditions. But it seems clear that in very many cases we may expect to find adaptive changes going on which fit the animal more accurately for that particular type of life it has always followed. It is difficult to believe that changes in the flora and fauna of the surface layers of the great oceans, of such a kind that entirely new sources of food have presented themselves, have occurred during the latter half of the Tertiary period which has seen the evolution of the whalebone whales. The diatoms and peridinians which form the foundation of all life in such waters were surely represented in Eocene seas, and it is very difficult to believe that copepods, or animals of similar habits, which form the next link in most oceanic food chains, did not occur as abundantly at that time as they do now. Indeed, although all the organisms living under such conditions may be expected to have changed during the Tertiary period, it is very difficult to believe that these changes, in effect of systematic position, are of any serious importance to the animals which prey on them. It would follow, then, that the evolution of oceanic or indeed of more nearly inshore surface-feeding animals may have gone on without any appreciable change in the general and feeding habits of the stocks in which they are seen to take place.

One group which is peculiarly fitted to give evidence on the matter is that of the plesiosaurs. These animals make their first appearance probably in the Rhaetic, but the first materials of any importance are found in the basal zone of the Lower Lias. From that time up to the Cenomanian period of the Cretaceous they are abundant; they were world-wide in their distribution; the bulk of them were marine; but from mid-Jurassic times to the Upper Cretaceous certain stocks had left the sea, and their remains are found

165

in fresh-water deposits. If we ignore these special members we find that the plesiosaurs as a whole are typically found in comparatively shallow-water marine deposits, but that they are represented even in the English chalk, which though not deposited at any extreme depth is at any rate considerably removed from a shore line. But the very wide distribution of the group, the fact that its members reached New Zealand, for example, suggests that they had the power of crossing wide seas. In the Lower Lias of England, and indeed generally, they lived in waters which supported large numbers of cephalopods, including belemnites and decapods without a guard (cuttlefish, in fact). These seas were inhabited by fish of all sizes, from leptolepids the size of a minnow to great sharks and sturgeons five or six feet long; and the fish present a wide variety of types, some with crushing dentition, others predaceous; and one group, the pachycormids, seems to have been as active and quick moving as the mackerel and swordfish of today. It is unfortunate that we still lack detailed descriptions of most of the earlier plesiosaur skeletons, but a great part of the material of this group is known to me personally, and many of the more obvious structural changes which go on during its existence can readily be seen.

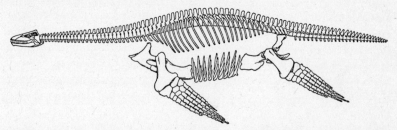

FIGURE 69. The skeleton of the plesiosaur *Cryptocleidus*, from the Oxford Clay of Peterborough, England. This figure is quite characteristic of a member of the group, belonging to neither of the two evolutionary lines discussed in Chapter VII. After Andrews.

The group divides itself into two, large-headed and small-headed plesiosaurs, living side by side from the Lower Lias to the Upper Cretaceous, and doing so apparently all over the world. Each of these divisions is itself complex, and to disentangle the various phyletic lines is at present impossible. All that can be done is to pick out from the whole series two stocks which it is possible to identify, chiefly on the characters of their cervical vertebrae, the selection being controlled by all other parts of the skeleton. The first group which I propose to consider is that which ends in the genus *Elasmosaurus;* the second—much rarer for individuals—leads on to *Polycotylus.*

The elasmosaur series, which is characterized by possessing cervical vertebrae whose centra are elongated and have very narrow zygapophyses, begins in the Lower Lias in an animal which has a neck containing thirty-five vertebrae, the largest number known in the Lower Lias. In the Upper Lias the group is represented by a form with thirty-nine or forty cervical vertebrae; by the Oxford Clay there are forty-two; in the Kimmeridgian the number is known to be high; while in the Cretaceous elasmosaurs the maximum of seventy-six is attained. In the initial form the neck is extremely stiff, the more posterior cervical vertebrae possessing wide, essentially flat faces closely opposed to one another; and the cervical ribs, rigidly attached to the centra, stretch backward parallel to the vertebral column, overlapping one another for long distances, the intercostal muscles being extremely short. Farther forward the neck is more flexible, but it seems evident from a consideration of the whole column that although the neck may have bent round through a considerable arc the snakelike flexibility represented in many restorations did not exist. This neck passes back through a sudden widening and deepening into the trunk, which is of considerable length, while the tail, apparently of nearly circular section, was no longer than the back, showing no signs of compression in any direction and no evidence of the existence of any fin. It is clear that such an animal is singularly badly designed for rapid movement through the water. It has a very large surface area in comparison with its bulk, the anterior part is slender, indeed essentially pointed at the front, the point of maximum depth is about two-thirds of the total length behind the anterior end, and over a long region there are concavities in any section passing through the axis of the animal. In other words it does not conform in any way to the requirements of that streamlined form exhibited by all rapidly moving aquatic vertebrates, whether they be fish, ichthyosaurs, or whales. The creature possessed large pectoral and pelvic paddles, quite inflexible, and so far as the evidence goes not externally divided into fingers or toes; it swam by rowing itself through the water, a mode which requires the existence of a power of feathering the paddles as they are moved forward through the water and rotating them through 90 degrees, so that the broad plane gets the grip on the water which is necessary for their effective use.

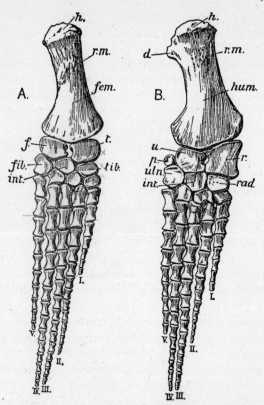

FIGURE 70. The paddles of the elasmosaur *Muraeno-saurus,* to illustrate the character of the proximal bones, with the attachments for the chief swimming muscles. The shortening of the bones of the fore-arm and leg is vividly shown. A, left hind paddle; B, left fore paddle from ventral side; *d.,* tuberosity of humerus; *f.,* fibula; *fem.,* femur; *fib.,* fibulare; *h.,* head of humerus and femur; *hum.,* humerus; *int.,* intermedium; *p.,* pisiform; *r.,* radius; *rad.,* radiale; *r.m.,* ridges for muscle attachment; *t.,* tibia; *tib.,* tibiale; *u.,* ulna; *uln.,* ulnare; I–V, first to fifth digits. After Andrews.

If, then, it is evident that these plesiosaurs are not designed for rapid movement in the water, it is interesting to speculate as to why they have developed such disproportionately long necks. The actual Triassic ancestors of the true plesiosaurs seem to be unknown, but it is evident that the nothosaurs are collateral ancestors, creatures derived from an early plesiosaur stock. The nothosaurs have limbs which still retain a flexibility between the femur and the tibia and fibula; and the tarsus and the toes, at any rate, seem to have been separately movable. They are in fact less fully adapted for an aquatic life than plesiosaurs. They are found abundantly in the shallow-water deposits of the German Muschelkalk and at various horizons in the Black Shales of Switzerland and north Italy. But all of them show to some extent an elongation of the neck; the number of vertebrae in the neck is twenty or more, greatly exceeding that found in any normal land-living reptile of the same time. The neck is variable in length; in *Nothosaurus* itself it may have twenty or twenty-three vertebrae, and other similar forms have similar proportions. Thus the elongated neck of the plesiosaurs is of old standing, and the conditions which led to its development must be investigated in early Triassic representatives of the order.

The only such animal known is *Trachelosaurus* from the Bunter, which Broili and Fischer regarded as an ancestor of the Sauropterygia. This animal possesses about twenty-one elongated cervical vertebrae; so far as is known there is no evidence of any great aquatic adaptation to be seen in the skeleton of the animal, which was found in an essentially terrestrial deposit at Bernburg. If correctly referred, i.e., if this animal really is a plesiosaur ancestor, it would suggest that the elongation of the neck seen in *Nothosaurus* and *Plesiosaurus* alike was developed either in the process of adaptation to aquatic life or even on land; that the nothosaurs and plesiosaurs inherited the condition from their ancestors, and having gone into the sea were thereby presented with certain opportunities which they then exploited. We may therefore begin our exposition with the fully developed *"Plesiosaurus" conybeari* and with those other forms which are associated with it.

We have seen that this animal is not well fitted for rapid progression through the water and have therefore to consider its possible feeding habits. The skull is eighteen inches in length, the lower jaw extending from end to end so that the possible gape was very wide. The dentition consists of an irregular series of long, rather slender, sharp-pointed teeth, the upper and lower dentition interlocking, so that the whole is capable of forming a highly effective trap. It is evident that it is well fitted to capture and retain fish, even large fish, and squids, and I have no doubt that this was indeed its diet, though there is no direct evidence on the matter. Throughout the series, from *"P." conybeari* through *Microcleidus, Muraenosaurus, Colym-*

bosaurus to the various typical elasmosaurs, there is a steady reduction in the relative size of the head and increase in the length of the neck, with not only the number of vertebrae increasing but the individual vertebrae becoming longer, while the back remains about the same length and the tail varies little but at any rate shows no active elongation. In other words the reptiles which are members of the series become steadily less and less well adapted to quick forward movement through the water. At the same time their paddles decrease in size; those of the latest forms—the elasmosaurs proper—though still of considerable size, are relatively smaller than those of *P. conybeari*, and indeed of the intermediate forms. Their decrease in size is shown most clearly in a shortening of the proximal elements, the humerus and femur.

The obvious interpretation of these strange facts is that the clasmosaurs did not capture their prey by chasing it through the water but must have adopted some other method, whose nature might be inferred from a consideration of the mechanics of the skeleton and its musculature as a whole. The first stage is therefore to see how far it is possible to reconstruct the musculature from a consideration of the visible traces of the insertion of the muscles on the bones, and on the evidence given from the actual shapes of the bones where they lie in contact with muscles. Fortunately the musculature of the reptilian forelimb and shoulder seems to have been essentially uniform in the more primitive members of most orders, so that examination of the nothosaur humerus, which still retains something of the shape of its land-living ancestor, may be expected to show muscle insertions which are identifiable and which, taken into consideration with a similar study of a shoulder girdle, should allow the whole of the musculature to be reconstructed with great probability. This can be done, and it is possible to show that the shoulder musculature of nothosaurs is a simple modification of that which is found in *Sphenodon* and in lizards, and also in primitive pelycosaurs of probably rather remote relationships.

From a nothosaur we can pass to a Lower Liassic plesiosaur and find, not surprisingly, that in favorable materials (which are very rare) it is possible to identify on the humerus all the muscle insertions to be found in *Nothosaurus*. From this conditions in later plesiosaurs can be determined, less completely but with considerable certainty. The shoulder girdle is well known throughout our series of long-necked plesiosaurs and, taken together with the known insertions of muscles on the humerus, enables the general structure to be determined without great difficulty. The humerus of a plesiosaur has an elongated shaft truncated by a hemispherical and cartilage-clad head, continued on the side of one broad surface of the paddle-shaped bone by an additional extension forming part of the same spherical surface. This head fits into a corresponding glenoid cavity

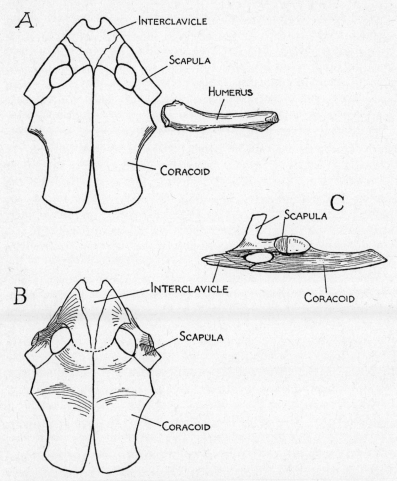

FIGURE 71. The shoulder girdle of *Plesiosaurus dolichodeirus*. Ventral (A), dorsal (B), and left lateral (C) aspects. To show the general character of the structure, and especially the distance between the ascending blade of the scapula and the glenoid cavity, the two points on which end the reference lines from the legend "scapula" in (C). This gap allows muscles to pass directly from the anterior part of the scapula to be inserted on the humerus so as to allow the paddle to be drawn forward in the swimming position.

divided almost equally between the scapula and the coracoid, and these two bones are expanded into broad sheets which stretch inward until they meet one another in a very powerful symphysis in the middle line. The thickness of the bone is greatest between the two glenoid cavities. The original dorsal blade of the scapula arises from a point considerably in front of the glenoid cavity and is directed upward and a little backward over the ribs. The remainder of the scapula, which is really an acromion, is, in all young individuals, a comparatively short sheet of bone stretching forward and inward toward its fellow; on its upper surface lies a thin sheet of bone, the clavicle, which is connected to its fellow by an interclavicle whose size varies greatly in different animals, apparently without any great systematic importance. As the animal becomes older the acromia extend farther forward and inward until they meet in suture, and ultimately grow backward until they reach and articulate with corresponding processes growing forward from the coracoids; as a result the elasmosaur shoulder girdle comes to present a ridiculous resemblance to a pelvis, with obdurator foramina between pubes and ischia.

An analysis of the musculature shows that the backward movement of the paddle in a swimming position, with the broad face vertical, is brought about by powerful muscles extending from the humerus to the coracoid and to the immensely powerful series of abdominal ribs which all plesiosaurs possess. Forward movement of the paddle is brought about by the contraction of the muscles which arise from the ventral surface of the clavicular arch, and of the acromion of the scapula, and by other muscles which pass forward and inward from the humerus behind the upwardly directed blade of the scapula to arise from the dorsal surface of the acromial part of the scapula. These muscles together enable the paddle to be drawn very powerfully backward during the swimming stroke and to be feathered and pulled forward in preparation for the next. Their action sets up stresses in the shoulder girdle and these all unite to produce a very powerful compression between the glenoid cavities, a stress which is met by the thickened bar of bone forming part of the coracoid which extends across the whole ventral part of the shoulder girdle and is necessarily retained even in *Elasmosaurus*, where the hinder parts of the coracoids are no longer in contact with one another. It can easily be seen that the changes in the shoulder girdle observed in the elasmosaur series are such as to decrease the size and efficiency of the angle of insertion of the swimming muscles on the forelimb, so that the proportionate reduction in size of this structure is comprehensible. But at the same time the muscles inserted on the clavicle and acromion may increase in size and gain a more effective angle of insertion on the humerus, so that it becomes possible to draw the front paddle forward as powerfully as it is pulled backward in the normal swimming stroke. It may be added that the arrangement of the musculature is such that the forward stroke of the paddle may be made with its broad face vertical so that it holds the water and

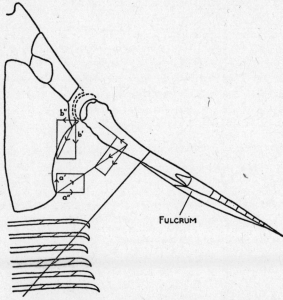

FIGURE 72. Diagram to show the resolution of the forces, resulting from the contraction of the coracobrachial muscle, into components at right angles to the principal plane and parallel to it. See text.

would, if both paddles are working in the same way, propel the animal backward. Additional evidence that this power of backing water is important in the later small-headed plesiosaurs is given by the fact that, while in some Lower Liassic plesiosaurs the paddle is actually spoon-shaped so as to hold the water more adequately during forward propulsion, in the later forms the two broad surfaces of the paddle are similar, and the paddle is equally efficacious during movement in either direction.

Examination of the pelvis and hind limb leads to the same conclusions. The pubes, from whose surfaces arise the muscles which

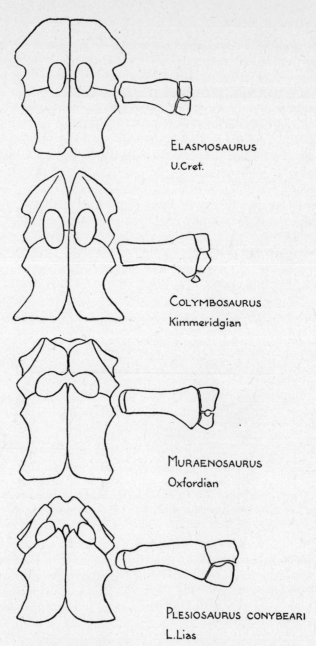

ELASMOSAURUS
U.Cret.

COLYMBOSAURUS
Kimmeridgian

MURAENOSAURUS
Oxfordian

PLESIOSAURUS CONYBEARI
L.Lias

FIGURE 73. Ventral aspect of the shoulder girdle of four
members of the elasmosaur series to show the reduction in
the coracoid, expansion of the anterior part of the scapula,
and shortening of the humerus which characterize the group.

174

draw the paddles forward, increase in area in the case of the elasmosaurs proportionally to the ischia, from which, and from the tail,
arise the real swimming muscles, and the hind paddles present a
reduction in over-all size similar to the pectoral limbs.

The significance of these facts seems to be quite clear. It is of
course absurd to believe that elasmosaurs could ever be called on to
proceed backward for any material time, but the changes would
enable the later members of the group to turn laterally quickly and
powerfully by going ahead with the one paddle and backing water
with the other. Such a power would enable a long-necked plesiosaur
to swing its head and the enormous neck behind it laterally through
a large arc and thus catch fish or cephalopods within its range without any movement of the body except such as might be imposed on
it by reaction from the neck movement. And the amount of such
movement could be reduced; indeed the whole body might be
swung round in the direction of the neck by backing water with the
paddles on one side and going ahead with those on the other. We
have, therefore, in these elasmosaurid plesiosaurs an example of an

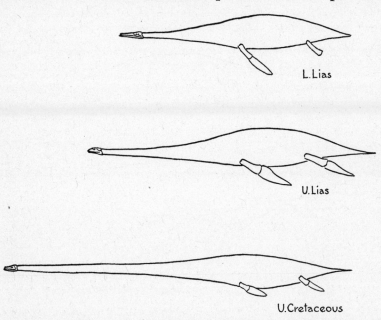

FIGURE 74. Outlines of the body form of three members of the elasmosaur
series. *Plesiosaurus conybeari* from the Lower Lias; *Microcleidus* from the
Upper Lias; and *Elasmosaurus* from the Upper Cretaceous; reduced to the
same "body" length.

adaptation to a particular type of feeding habit retained without modification through the many generations which separate the Cenomanian from the Lower Liassic members.

If this story be true we should expect to find that the large-headed, short-necked plesiosaurs exhibit very different changes from a structure which, in the Lower Lias, is not really very dissimilar to that of the long-necked forms. Unfortunately the large-headed plesiosaurs are less abundant and less adequately represented than those of the other great group. But a series of forms, *Sthenarosaurus* from the Upper Lias, *Peloneustes* from the Oxfordian, and *Trinacromerum* from the Upper Cretaceous seem to be essentially members of one stock. They show a series of changes in all parts of the skeleton of such a kind that the animal seems to become more and more nearly streamlined, though it never obtains the perfection of an ichthyosaur

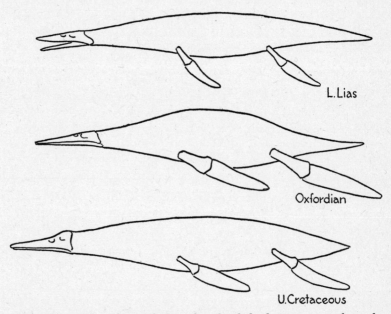

FIGURE 75. Body outlines of three long-headed plesiosaurs, to show the slight changes in shape. *"Plesiosaurus" rostratus*, Lower Lias; *Peloneustes philarchus*, Oxford Clay; *Polycotylus osborni*, Upper Cretaceous.

or a whale in this respect. The paddles on the whole become larger and the neck shorter. In the shoulder girdle there is no such enlargement of that anterior part of the ventral surface formed by the acromion processes of the scapulae and the clavicular arch as is seen so clearly in the long-necked forms, but the coracoids expand enormously in length, breadth, and area, and the swimming muscles, which have their origin from them, must have increased very greatly in bulk. In fact the large-headed forms show no adaptation to quick turning or to wide lateral sweeps of the head; they must have caught their food by running it down, and all the changes in their structure have the effect of securing increased speed through the water.

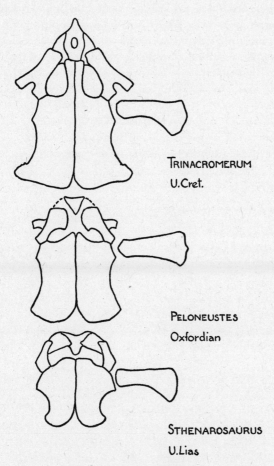

TRINACROMERUM
U.Cret.

PELONEUSTES
Oxfordian

STHENAROSAURUS
U.Lias

FIGURE 76. The shoulder girdles of three large-headed plesiosaurs to show the steady expansion of the coracoids, which give origin to the muscles used in forward swimming, and the lack of important increases in the anterior part of the girdle.

It follows therefore that, like the long-necked elasmosaurs, they kept unaltered through a vast period of time a series of feeding habits peculiar to them. Elasmosaurs and large-headed plesiosaurs lived side by side throughout the whole of their history. There was nothing in their environment to limit their choice of food; an early long-necked form could probably eat quite large fishes and an early short-necked plesiosaur could no doubt have pursued and eaten the smaller prey to which the long-necked forms appear to have restricted their attention. In other words the feeding habits of the animals must have been inherited from generation to generation without change other than that imposed by the evolutionary changes of the animals on which they feed. Habits of this kind, instinctive behavior in the true sense, must thus be determined by the hereditary mechanism, and may be expected to persist as long as the structural features which owe their appearance in each individual to that same mechanism. Such a persistence of habit is necessary if we are to believe that adaptive characters fitting an animal for some specific mode of life have arisen under the control of natural selection.

APPENDIX

Since this lecture was delivered two valuable papers on plesiosaurs have been published. The first by T. E. White contains a general classification of the whole group. Unfortunately Dr. White has not had the opportunity of examining the incompletely described early Jurassic plesiosaurs, and in consequence it seems quite evident to any one who has personal knowledge of the materials that his classification, founded on the character of the shoulder girdle, is unrealistic. It brings together animals which are, on a totality of characters, extremely dissimilar, and separates others which are probably closely allied.

Welles (1943) gives a description of some elasmosaur skeletons which are certainly the best hitherto collected, but some of his reconstructions of shoulder girdles seem to me entirely unjustified; for example, the figure of the restored shoulder girdle of *Thalassomedon hanningtoni* on Pl. 23 is mechanically entirely unsound. It does not agree with the drawing of the same structure in Fig. 14, which is no doubt correct in the contact between the anterior processes of the scapulae. But the position of the clavicular arch in particular is quite impossible. As it is drawn, this structure is composed of three bones: an interclavicle with a deep keel, with clavicles fused onto the visceral surface. The whole structure is 60 cm. across and it is represented as overlapping the extreme anterior extremity of the scapula by an average distance of 6 cm., yet the bone is an important muscle insertion and in addition serves to maintain the shape of the ventral surface of the whole shoulder girdle. In all plesiosaurs in which the structure is properly known the contact between the clavicle, even if much reduced, and the visceral surface of the acromial part of the scapula is very extensive indeed, the tip of the clavicle usually turning dorsally onto the root of the ascending part of the scapula. I have no doubt that this was in fact the condition in *Thalassomedon,* and also in *Aphrosaurus* and *Morenosaurus,* the other elasmosaurs described by Welles in which the structure is known. Indeed nothing in these papers leads me to alter the conclusions set out in Chapter VII.

I am quite unconvinced by Mr. Welles's discussion in his paper of the possible movements of the vertebral column and neck. Examination of naturally articulated specimens in which, except for its extreme anterior part, the neck is often perfectly straight and never, I think, shows any sign of multiple flexure, and comparison of such skeletons with well-preserved isolated bones have long led me to believe that the posterior part of the neck of long-necked plesiosaurs was effectively quite stiff and the possible movements in the midregion were small, though it seems clear that the first half dozen or so vertebrae were a region of flexibility. In such early long-necked plesiosaurs as retain two-headed cervical ribs, as for example in *Microcleidus,* these ribs throughout the posterior portion of the neck are backwardly directed. They lie one above the other so closely placed that the intercostal muscles were extraordinarily short, a matter of a centimeter or so, and the possible range of their contraction so limited that the range of their action can only have been exceedingly small and any extensive bending of the neck quite impossible. In *Muraenosaurus,* however, in which the cervical ribs are attached to the centra only by a single head, the ribs, as figured by Andrews, do not overlap at all, and there is in principle the possibility of considerably more movement. Nonetheless mere inspection suggests that the neck was exceedingly stiff.

How extraordinarily inefficient for rapid movement through the water *Elasmosaurus* must have been is well brought out by contrasting it with the ichthyosaurs. In Fig. 77 I have drawn in outline three ichthyosaurs. The first, the Triassic *Cymbospondylus,* is restored from the skeletons prepared by J. C. Merriam. The next, *Ichthyosaurus*

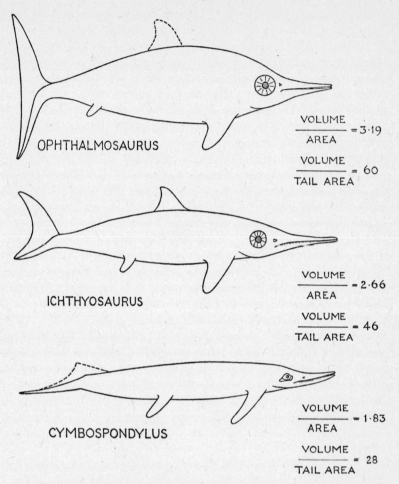

FIGURE 77. Restored outlines of three ichthyosaurs, *Cymbospondylus,* Upper Trias, California; *Ichthyosaurus,* Upper Lias, Württemberg; *Ophthalmosaurus,* Oxfordian, England. All are reduced to the same length, and their origin is described on pp. 179–180.

quadriscissus, from the Upper Lias, is traced from a photograph of one of Dr. Hauff's beautiful specimens showing the outline of the fin. The third, *Ophthalmosaurus,* from the Oxford Clay, is founded on the mounted skeleton in the British Museum, which is extraordinarily well preserved; the tail fin is taken directly from that in the museum at Munich, which comes from the Solnhofen slate. This tail fits quite accurately the downturned posterior extremity of the

Ophthalmosaurus specimen. The dorsal fin is unknown; the pectoral and pelvic limbs are taken merely from the skeleton, their outline being hypothetical, as are of course those of *Cymbospondylus.*

I have calculated the volume of the body, the total surface area, and the area of the tail fin for each of these animals, and find that, other things being equal, the volume divided by the surface area is 1.83 in *Cymbospondylus,* 2.66 in the ichthyosaur, and 3.19 in *Ophthalmosaurus.* The surface area measures the amount of frictional resistance to movement; the volume is a guide to the amount of energy which the animal's musculature is capable of expending over a given period. This would imply that the ophthalmosaur is capable of higher speed than the ichthyosaur, and the ichthyosaur than *Cymbospondylus.*

In addition resistance to movement is afforded by the production of turbulence in the water in contact with the body, resulting from inadequate streamlining. Mere inspection will show that *Cymbospondylus* is very imperfectly streamlined, *Ophthalmosaurus* much the best, so that the differences in speed may have been very considerable. It is therefore interesting to note that the tail area per unit weight of body becomes smaller between the Trias and the Oxford Clay, a change which presumably reflects an alteration in the character of the swimming movements similar to that which distinguishes a dogfish from a salmon. The paired limbs in ichthyosaurs seem to have functioned generally as stabilizers. These, used as paddles, as they clearly could be, may have been important in producing sudden changes of course.

It will be noted that while the long-necked elasmosaurs depart more and more widely from the typical streamlined form of *Ophthalmosaurus,* the large-headed pleisosaurs make a nearer approach with time, though very evidently they were propelled by the movements of the limbs and not of the tail. Such a comparison with ichthyosaurs makes it quite evident that the long-necked plesiosaurs caught their prey in some quite exceptional manner such as that which I have suggested.

VIII. POSSIBLE MACHINERY

IN THE PRECEDING CHAPTERS I have shown, by the consideration of actual cases, that it is possible to examine the structure of an animal and, by an analysis of the conditions to which it is or may have been subjected, to demonstrate that some of its characters depend on what is called adaptation to its own special mode of life. Thus the stiff paddles of all plesiosaurian reptiles are there because the animals used them in order to swim actively in water. The fact that in the elasmosaurid plesiosaurs the two broad surfaces of these paddles are alike is related to the fact that in them the power of "backing water" was much used, and the high development of this power depends on the structural arrangement of the bones and muscles of the shoulder.

These structural qualities are thus intimately bound up with one another, and may be brought into relationship with the characters of the small skull and immensely long neck which these animals possessed. The whole structure becomes intelligible on mechanical grounds if the elasmosaurs caught active, quick-moving prey (fishes or cephalopods in all probability) not by running them down by sheer forward speed but by rapid flicks of the head. This was made possible by movements of the great neck whose anterior part had a lateral flexibility, helped by a turn of the animal as a whole brought about by backing water with the paddles of one side while those of the opposite side of the body were pulling forward. An analysis of this kind, by the application of ordinary mechanical principles to the understanding of the locomotory apparatus, can be made of every animal, and actual experiment may show that its conclusions are correct.

If we push such an analysis of stresses as far on as it will go we shall find unexplained a residue of parts which are, as it were, the raw material from which the structure whose mechanical qualities we have investigated was built up. Thus in the case of the elasmosaurs we have the various bones (the scapula, coracoid, and so on) of the shoulder girdle and forelimb, and the muscles attached to them, which, by their well-regulated contractions, bring about the animal's movements. By a comparison of two very different kinds of plesiosaurs we discover that these parts are common to both, that in fact they represent the essence of the structure of plesiosaurs in general.

The whole argument is a characteristic example of morphological method. As I showed in the first chapter it is possible by such methods to make predictions, to demonstrate by analysis so carried out over a very much wider field that animals are possible (and may have existed) which possessed definite describable qualities not to be found in any living or fully described fossil form but probable because they conform to morphological requirements of the great group of which they would be members. I have shown in three cases that such hypothetical animals exist. The Aphetohyoidea are a group of fishlike vertebrates in which the predicted condition of a spiracular gill slit of full size and function actually existed. Professor Stensiö has shown that the very ancient *Cephalaspis* is a vertebrate in which the predicted existence of an open functional gill slit lying in front of the mandibular arch is actually realized. The cynodonts are reptiles in which the tympanic membrane is borne by a bone of the lower jaw; and the homologues of the outer bones of the chain of auditory ossicles of mammals, the incus and malleus, are the quadrate and part of the lower jaw, respectively, in a way long predicted by Reichert. Thus the validity of morphological method may be established by its power to make verifiable predictions.

I have shown that by applying such methods it is possible to sort out the remains of fossil vertebrates into groups, large or small, which are the classificatory elements of systematics. We then take all the fossil animals included within a small group and arrange them in their order of appearance in time, using the justifiable methods of the stratigraphical geologist in order to do so. Mere inspection of such series usually shows that all the later members, no matter how they differ from one another in obvious qualities, agree in differing from all the earlier in definite ways. If now, by the application of morphological methods to smaller detail, we narrow the width of our group of forms, we shall find that we ultimately get a series of animals showing a gradual change in structure in those ways which distinguish all early from all late members of the group, which is as continuous as it can be in view of the limited number of individuals available to us.

It is obvious that the only intelligible explanation of such a series of structural changes is that it depends on the evolution of the later from the earlier members of the group. But it is further obvious, and is of the highest importance, that the course of evolutionary change of these essential qualities must have been the same in all allied stocks, for were this not the case it would be impossible for all the late members of so diverse a group as the Labyrinthodonta, for example, to agree with one another in those ways in which they all differ from all the earliest members of this group.

All modern paleontological work, on invertebrate groups as well as on the various classes of vertebrates, has shown that a parallel evolution of this kind is so widespread through the animal kingdom that it is the general rule, and that the discovery of the mechanism on which it depends should be one of the chief aims of the student of theoretical biology. Parallel evolution of this kind is not, however, universal. The evolutionary history of the plesiosaurs, which I discussed in Chapter VII, is entirely different in that the small-headed elasmosaur line steadily diverges more and more widely from the large-headed group which ends in *Trinacromerum*. The initial members of the two groups, which lived side by side in the Lower Liassic seas of the south of England, are not very different in structure, while the ultimate results of the evolution are extremely dissimilar.

The elasmosaur series, which is the more completely known, shows well that the evolutionary changes which go on proceed regularly with time, at any rate to the extent that each known member of the series differs, in respect to the qualities which distinguish all the later from all the earlier members of the group, from its immediate predecessor in exactly the same ways as its immediate successor does from it. In fact the course of evolution is direct, as if designed to secure a definite final result; it is orthogenic in Osborn's use of the term, even if not in the original intention of Eimer.

This consistency of the direction of evolution throughout a long period of the history of a group is a widespread phenomenon; it exists in the labyrinthodont groups, which have a history of parallel evolution, and in those mammal-like reptiles in which the parallelism of change is less absolute, as clearly and precisely as in the divergently evolving plesiosaurs. It is therefore necessary to discuss its significance separately from that of parallel evolution.

Even the most detailed discrimination of separate stocks only allows us to reduce our groups of labyrinthodonts to belts which contain a single genus at any one horizon, this genus itself including usually more than one and often a considerable number of species. The materials with which I have worked are such that the problem of the discrimination of species and the discussion of the nature of specific differentia scarcely arise. Nevertheless it is difficult to avoid some treatment of the matter, because it may well be held that the species is the only true entity (other than the individual) existing within the animal kingdom, all other systematic groupings being the product of the accepted procedure of taxonomists. Although in practice "a good species is that recognized by a competent systematist," in fact it is probably in all cases a group of individuals, each in principle capable of producing fertile offspring by breeding with any other individual of the opposite sex within the species. Furthermore it is a necessary feature of a species that its members do not under natural conditions breed with members of other species, or if they do the offspring are not fertile.

It is necessary to explain the significance of this absence of interspecific breeding. It seems clear that to some extent, though not entirely, the morphological differences between two species of the same genus may be expected to depend on a difference in the sum of all the genes existing in their chromosomes, that in fact the two differ because homologous genes are allelomorphs in greater or less number. This is an assumption at present, even in the best investigated cases, as for example in the genus *Drosophila;* it has not been shown that the specific differences which exist, whether of structure or of physiology, can be accounted for by any number of mutant genes, but it is a reasonable assumption.

The geneticist can and does build up individual specimens of *Drosophila* which contain many mutant genes and hence differ greatly in structure and appearance from the normal wild type. But these may be, and generally are, fertile with normal individuals. We can thus determine the character of the population which would result from the interbreeding of the complex mutant and wild type. In such a population every possible combination will be present in calculable numbers. Thus, unless some factor is present which will prevent the mutant individuals from breeding with the normals, the group of mutant genes in an individual which causes the difference in structure will fall apart again and no specific difference will be established.

It is immaterial how this inhibition of breeding is brought about. The classical geographical isolation envisaged by Darwin probably often occurs and may exist even in continuous land areas if the animals under consideration are restricted to some relatively unusual ecological niche, so that the colonies are sufficiently isolated to render it difficult for individuals to wander from one to another. In these circumstances it is obvious that if the groups of initial individuals which founded colonies differed from one another in the genes which they contained, then after a period sufficiently long to have allowed the establishment of a steady state the populations derived from them will differ in the distribution of their variability, and perhaps even in the mode, for each measurable character. It is not improbable that many of the geographical races recognized within species have such an accidental origin. Indeed it is evident that geographical races of man, as of other animals, have only such a statistical significance; identical individuals may and do occur in different geographical races, which can thus only be distinguished from one another by the measurement of numbers of individuals of each race.

Even where individuals of two species live side by side, as for example on the sea floor or in a rock pool, and are potentially capable of breeding together, they may in fact never do so. It is often found that marine animals have very restricted breeding periods, when the eggs and sperm of a large proportion of the individuals of a single species in an area are ripe and are shed into the sea. In many such cases the breeding period may only extend over a few days, perhaps only a single day at the time of a new moon or of a full moon. In such cases it is clear that if two species of the same genus shed their gametes, one at new moon and the other at full moon, they will never in fact produce interspecific crosses, even if there is no positive block to effective fertilization. To such cases of isolation by difference of breeding season may be added those which arise from mechanical nonadaptation of the pairing apparatus or from a lack of the necessary pairing instinct.

But perhaps even more important are those cases in which there is a genuine sterility, either of the initial pairing or of the first generation, resulting from such an interspecific cross. There are cases in which either mechanical or chemical difficulties prevent the perforation of the egg by the foreign spermatozoon. Even if a foreign sperm penetrates an egg and thereby activates it to cleavage, it is by no means certain that a true fusion of the nuclei will occur, or that the joint nucleus, if formed, will be normally divided at mitosis. If it is so formed and carried on to the primordial germ cells it is likely that the elaborate process of meiosis, with its necessary accurate pairing of homologous chromosomes, will prove impossible, or at least so hazardous that few fertile gametes will be formed. The recent very rapid advances in the study of chromosomes have shown several ways in which such difficulties in meiosis may arise.

The number of chromosomes which occurs in the nucleus of an egg or a spermatozoon is, except for the existence of a sex-determining mechanism, constant in every individual of a species. This haploid number is often found to be the same for several or all the species of a genus. But in some cases some one species or group of species differs from its fellows within a genus by having a haploid number greater by one or more chromosomes. The obvious explanation is that this chromosome has either appeared in the nucleus as a duplicate of one of the others carried in at some abnormal mitosis or represents a fragment of one of the original chromosomes which has gained a new "spindle attachment." However it has arisen, the existence of such an additional chromosome is bound to produce difficulties if, by a cross between two individuals of different species, it comes to form part of the complement of a nucleus of a hybrid. When such a hybrid forms its sex cells, at a stage when homologous chromosomes lie alongside one another, so accurately placed that each individual gene in one is attached to the equivalent gene in the other, the newly introduced chromosome will, if it is a mere duplicate, have no fellow to pair with; while if it is part of one of the original chromosomes it will in general be unable to pair with the equivalent portion of its homologue, even though the pairing of this chromosome with the other of the two chromosomes into which it has been supposed to have divided leaves vacant a region appropriate to it.

Chromosomes are now known to suffer by accident internal rearrangements which hinder pairing. Cases are known in which a short length of a chromosome has in some way become detached and reinserted in its old position but inverted, so that within the region of the inversion the genes retain their correct positions with respect to one another but their order is inverted with respect to the rest of the chromosome. In other cases a short length of chromosome may fall out and be reinserted in a new position. Sometimes a short length of a chromosome is actually lost, "deleted," or it may become attached to another chromosome.

Any one of these changes, all of which occur spontaneously in *Drosophila* cultures, may lead to sterility, partial or complete, between the abnormal individual and normal animals, but no one of them will necessarily interfere with the fertility of two animals both of which possess the quality so acquired. But such rearrangements of genes, without any actual mutation or modification of any of the genes themselves, lead to change in the phenotype, the character of the individual in whose cells they have taken place.

Thus if the interpretation of the essential nature of a species which I have given above is accepted, the group of individuals whose nuclei have the newly acquired rearrangements of genes is a species, because these individuals are fertile among themselves, breed true, and are sterile with members of the species from which they have arisen, and may be morphologically separable from it.

Until a few years ago most geneticists and many zoologists imagined that species of the same genus differed structurally and physiologically from one another because of the different gene complements contained in their respective nuclei; that the differences between a rat and a mouse, for example, would be explicable on the basis that different allelomorphs of few or many genes were present in them. There seems at present to be no well-investigated case in which such an explanation is applicable, though the results of hybridizing pheasants suggest that in these birds, at any rate, some part of the morphological differences between species may be so explained.

The conception of the nature of species which I have just expounded suffers, when regarded as a definition, from the disadvantage that it is inapplicable to fossil or indeed to dead individuals. But it is possible to get out of the difficulty by saying that dead individuals belong to that species with whose members they would, had their span of reproductive life been extended, have been capable of breeding.

I have entered into this short and much condensed discussion of the restricted problem of the evolution of species in order to bring out the limitations of fossil materials. It is evident that the paleontologist can never discover whether or not the members of two apparent species actually bred together, even if their remains occur side by side in the same bed of rock. He is incapable of observing ecological isolation, isolation by difference of breeding season, by chromosomal differences, by failure of pairing instinct, or otherwise. And similar disabilities prevent his forming any views on the nature of heredity. He is therefore compelled to accept the conclusions of geneticists, and other students of biology, whose materials do yield direct and valid evidence on such matters; and it is methodologically improper for him to make any postulates on such matters which cannot be confirmed by observation and experiment with living material.

The nature of such limitation is best illustrated by example. During recent years both vertebrate and invertebrate paleontologists have suggested that the difficulty which always exists of sorting out actual lines of descent—essentially of species from species, "lineages," from any large material representing the history of a family, or smaller group, in a short space of time in a small area—depends on the common occurrence of interspecific crossing, the lineages each supposed to be composed of evolutionary series restricted at any one moment to a single species, anastomosing with one another and then re-separating with new qualities acquired by hybridity; that in fact the evolutionary lines are not independent, parallel, convergent, or divergent as the case may be, but actually form a network. Such an interpretation of the evidence requires assumptions about crossing of a kind which, as I have just shown, can never be substantiated from the examination of fossil materials, and which hence can justifiably be made only if it can be shown to occur sufficiently frequently today. In fact the occurrence of such interspecific crossing with complete fertility in the first and subsequent generations of hybrids has never been shown to occur at all in nature among animals, even in those cases, such as coral reefs, when the conditions seem to render it most likely. The assumption is thus illegitimate, and the suggested phylogenies may be ignored until some confirmation of the basal assumption is obtained from the only source possible, that is, from the observation of living animals in their natural habitat.

The phylogenies with which I have been concerned in this book do not suffer from such disabilities; they represent the evolutionary changes in relatively important structures which occur in large groups of the animal kingdom. The individual animals involved are allotted to their groups by structural similarities and differences, determined by observations directed by ordinary morphological reasoning. And the methods of morphology are shown to be valid because they enable us to make verifiable predictions.

It is convenient to begin with a consideration of the case of the plesiosaurs, whose story was recounted in the last chapter. Here we trace the gradual development of very dissimilar animals of Upper Cretaceous age from very similar ancestors found in the Lower Lias. In neither of the two contrasted series, the small-headed elasmosaur and the large-headed *Polycotylus,* do we possess an actual lineage; in each we have usually more than one species within the genus at each horizon, but the series as a whole presents a connected story of the course of evolutionary change.

By reconstructing the musculature of these animals on the basis of the recognizable traces of the insertions of the individual muscles on the bones, and by studying the architecture of the bones themselves, one can determine the possible motions of the body and its paddles, and in this way show that the two extreme forms of Upper Cretaceous age are adapted to different forms of feeding. Thus the main evolutionary changes are typical cases of adaptation to specific habits and modes of life.

Two explanations only have been given of the mechanism of such adaptive evolution, the Lamarckian and the Darwinian. The Lamarckian explanation requires the "inheritance of acquired characters," clearly a phenomenon which, if it exists, can never be recognized in fossil material, and can only be used by paleontologists as a working hypothesis if it is acceptable to zoologists who have examined it experimentally. In my opinion none of those cases in which it has been held to occur can rightly be accepted as valid evidence; they are all inadequately controlled or deal with a phenomenon of a different type or are subject to grave doubt.

The classical Darwinian explanation of natural selection is also beyond the examination of the paleontologist, but it is accessible to zoologists and has indeed been increasingly explored in the past few years. The Darwinian postulates are only a) that those individuals of a species which are better fitted to their conditions of life will leave more offspring than those which are less well adapted; b) that the offspring of such well-fitted forms will inherit to some extent the good qualities of their parent or parents.

The most obvious reason why well-fitted forms should leave more offspring than others is that the length of their breeding life is longer. And this should be capable of being established by observation. Consideration of the kind of evidence required will show that it will be exceedingly difficult to obtain.

It is clear that the whole conception of natural selection is a statistical one; it is valid, if at all, only for populations and not for individuals. The existence of a selective death rate can thus be established only by knowledge of a large population from the time when its members begin to breed to the time when all, or at least a large proportion of them, are dead. And the observations must be made without so disturbing the conditions that they become no longer natural.

One example of a case in which it might be possible to secure evidence may be given in broad outline. The herring fisheries of the

Atlantic coast of Europe are based on shoals of fish which have gathered for breeding, and the mode of fishing by drift nets or trawls is one which is probably nonselective, so that the catches, which are enormous, even relative to the total population, may be expected to give a true sample. It is possible to determine the age, that is, the year of birth, of each fish by examination of scales; and the length of the fish at the end of each year of life can be calculated. The very prolonged investigations carried out under the coordination of the Conseil International pour l'Exploration du Mer have determined the breeding places and plan of migrations of most groups of herring. Thus by determining the constitution of the population of herring of one definite year class from the time when they are three or four years old and begin to breed until they vanish from the catches some five years later, it might be possible to show that herring with some special qualities formed a larger and larger proportion of their year class as this was followed from season to season. And if it could then be shown that the assumed facts could not be accounted for by immigration of newcomers we should have the evidence we need. But in fact these observations have apparently produced no evidence of a selective death rate. Nor was Professor Schmidt able to find any evidence from his observations on *Zoarces,* the viviparous blenny, under the very favorable conditions of observation in the Roskilde Fiord.

In the case of land animals the possibilities of observation are much less. The types of observation needed are very much the same as those which must be made of agricultural crops in field experiments, and in this case we can estimate the probable error of the experiments and discover that even in the best designed trials the error is of the order of 5 per cent, an amount in all probability very much greater than any due to a selective death rate of normal individuals of a species in nature. Thus the probability that we shall ever be able to observe such a selective death rate in nature directly is very small. But it has recently become clear that indirectly we can establish its existence.

If we can find an isolated population of a species, for example of *Drosophila,* and from this population collect females which have already paired, we can induce these samples to lay eggs in captivity and rear a generation of flies which, had their mother not been caught, would have formed part of the original population. By crossing these individuals among themselves we can obtain a second generation in which any mutant recessive genes which were con-

tained in the original wild pair of flies will become obvious as phenotypes. If we deal with sufficiently large numbers of wild flies we
can determine the extent to which wild flies are heterozygous for
some mutant genes. If we know the percentage of individuals in the
population which contain a definite mutant gene we can calculate
the number of phenotypes exhibiting this mutation which will appear in the population after it has reached a steady state, provided
we assume that there is no preferential mating and no difference in
the viability, under natural conditions, of the wild type and mutant
individuals and the heterozygotes between them.

Such observations have been made by Timofeev Resovski and by
C. Gordon, who both found that recessive mutations are commonly
found in wild individuals; Gordon obtained six different mutations in
a small number (twelve) of specimens of *Drosophila subobscura.*
Indeed nearly every other fly caught carried at least one recessive
mutant gene. It is obvious from the figures that the number of double
recessive individuals in which mutation is actually visible in the
phenotype occurring in the population is very greatly below expectation, a result only to be accounted for by a high death rate of these
types.

Professor R. A. Fisher, by an analogous but more indirect reasoning applied to Professor Nabour's grouse locusts (*Gryllotetrix*), has
shown that there is also satisfactory evidence of a greater mortality,
or at least of a lower effective fertility, of certain mutant forms. Thus
there is real evidence that the differential death rate, or differential
fertility, which is the first of Darwin's postulates, does actually occur
in nature.

In a series of papers, the first published in 1927, J. B. S. Haldane
has calculated the effect of a selection imposed on a population in
which a proportion of the individuals possess a mutation which exhibits a definite mode of inheritance, dominant or recessive, with a
gene in an autosome or in a sex chromosome. The whole series
of investigations shows that the effect of even a very small selective
advantage, even one-tenth of one per cent, is unexpectedly great in
what is, from a geological standpoint, a small number of generations.

It thus appears that, so far as the evidence goes—and at present
it relates only to insects—natural selection is a real phenomenon,
and that it may be accepted and used by paleontologists in the interpretation of their phylogenies.

Thus the evolution of the two divergent groups of plesiosaurs can
be explained by a strict Darwinian interpretation. But such an ex-

planation, or indeed any other which is concerned with the evolution of adaptations, can hold only if two postulates be granted, for
they cannot be established by observation. These are, first, that the
feeding habits of all the members of each line remain the same
throughout the whole life of the stock; and second, that the food
available and the general conditions of life either remain uniform
or change, if they do so, in a definite direction where life is not
rendered easier for the animals in question.

The plesiosaurs which we have considered belong to groups whose
latest members have a world-wide distribution. Elasmosaurs, for
example, occur in Kansas, Russia, England, Queensland (Australia),
and New Zealand; and *Polycotylus*, or animals very similar to it, occurs in the same localities. The distribution of these animals is thus
similar to that of the living Cetacea and seems to be most readily
understood if they were inhabitants of the high seas, independent of
shallow waters and the neighborhood of coasts; although if they
were attracted by food they could and did approach lands, so that
their remains may be found in such deposits as the English gault.

It seems evident that the general conditions of life of a surface-
living animal in the great oceans must always have been very much
the same; indeed any large, active swimmer could always have
sought for and found any environment which suited it, and might
even, like the great whalebone whales, have carried out an annual
migration to overcome any seasonal change in its food supply.

The food of the elasmosaur line was certainly fish, and probably
also cephalopods, and throughout its history both were present,
holostean ganoids in Jurassic times changing to teleost fishes in the
Upper Cretaceous, only a little before the disappearance of the
plesiosaurs. The large-headed *Polycotylus* group presumably fed
largely on other aquatic reptiles, ichthyosaurs and plesiosaurs
throughout their history, crocodiles from the Upper Lias to the end
of Jurassic time, mosasaurs and marine turtles in the Cretaceous.
It is difficult to believe that the position of their prey in a zoological
classification had any meaning to the predators, which were concerned only with its habits, the depth below the surface at which it
swam, its speed, and the rapidity with which it could maneuver. The
only evidence there could be as to the persistence of feeding habits
would be the nature of the contents of the gut, a matter about which
we have at present little information. It therefore appears that we
are justified in adopting a Darwinian explanation, the action of natural
selection, for the evolution of these groups of plesiosaurs.

One of the most striking features of these phylogenies is that the evolution throughout follows a direct course, not only for the animal as a whole but for all the individual regions of the body which have been studied. This constancy of direction of evolutionary change with time is a very widespread phenomenon, found in innumerable groups of vertebrates and invertebrates, and is one which could not have been discovered except by the investigation of fossil materials. It is important to realize that in many cases, in addition to that which we have just considered, it leads to adaptation to special modes of life. It is therefore important to consider other cases of direct evolution which seem to be nonadaptive and to discuss the mechanism which may have brought it about.

One such case, remarkable for its scale, is that of the evolution of the amphibia. In Chapter II I showed that in the order Labyrinthodonta the structure of every part of the head skeleton changes steadily and consistently with time, maintaining a constancy of direction during the very long period extending from the Lower Carboniferous (Mississippian) to the top of the Trias. And this constant direction of evolutionary change is unaffected by the transformation of animals which were essentially aquatic throughout their life to terrestrial forms which lived in water only during their larval life, and by the subsequent event whereby these land-living amphibia gave rise to secondarily aquatic descendants which, like their remote ancestors, lived in water for the whole of their lives. It is clearly impossible that these structural changes can be adaptive in the sense of securing better fittedness for some particular habit.

In Chapter III an analysis of the structural changes which go on in labyrinthodonts demonstrated that all of them could be explained as sequels of two more fundamental types of changes: a progressive dorsoventral flattening of the head and later of the whole anterior end of the animal; and a steady reduction in the amount of cartilage bone in the skeleton of the full-grown animal. These factors so greatly alter the skull shape, and result in so great a reduction in its mechanical strength, that many consequential rearrangements are necessary to allow the mouth to be opened as widely as is necessary, to house the eyes, and to give to the whole head the girder strength it must have in order to stand the very large stresses it is subjected to by the action of the powerful jaw muscles.

These consequential changes are thus adaptive in that they alone enable the animal to function in its new form. But the factors which have called them into existence are not environmental ones, they are not analogous to those which are responsible for the evolution of the plesiosaurs, but arise within the animal itself from some cause still unknown, which is at any rate not directly determined by any discoverable feature in its surroundings, either living or nonliving.

In Chapter IV I showed that, so far as the available material allows one to judge, nearly all the orders of amphibia pursue parallel courses of evolutionary change. In all of these the fundamental flattening and reduction of cartilage ossification produce spatial and mechanical difficulties which are met in very similar ways.

It seems possible to account for the secondary or consequential structural changes by natural selection. But any such mechanism leaves unexplained the remarkable similarity which exists in the actual form that the adaptation takes in many allied stocks. It is difficult to believe that no other arrangement of bones would have given the necessary strength, or allowed the mouth to have been opened sufficiently far, or met all the various needs as well as that which was in fact adopted. We are driven back to the conclusion that the possibilities of change in amphibia are limited in some way which in fact excludes other possible forms of change.

It is now certain that the greater part at any rate of the mechanism of heredity is contained in the chromosomes, and that while much of it lies in the genes these bodies can no longer be regarded as particles whose positions with respect to one another are without significance; the effect of any gene in influencing the patent characters of the individual of which it forms part is varied by altering its position within a chromosome among its linearly arranged neighbors. Despite these "position effects" it seems evident that the influence of mutation of genes has an overwhelming effect.

It seems probable that in many cases what are in adult animals single structures or functional activities may owe their quantitative similarity in different individuals to different sets of genes. For example, the daily experience of breeders makes it certain that the extent of the milk yield of a cow is largely determined by its heredity. The range of output for a single lactation may vary in normal healthy animals from less than three hundred gallons in the beef breeds to more than four thousand gallons in a particular cow now living in England, a case of genius in an unexpected direction. The average output in ordinary commercial herds cannot be closely estimated, but it is probably of the order of five hundred gallons in England. Nevertheless there are many herds averaging one thousand gallons, and some giving materially more. The very wide range of these figures, their general character, and the experience of breeders that bulls of milky ancestry do not "nick" with all good cows so as to produce high milk-yielding heifers seem to show that far more genes which raise the milk yield exist than are necessary to produce even such a yield as one thousand gallons; that there are, in fact, so many that it may be conceivable that cows of such productivity exist which have few "milk genes" in common.

This apparent multiplicity of separate genes capable of combining so as to give an additive effect, only a proportion being normally functional in any one individual, occurs in very highly developed domesticated animals, and it would be extraordinarily interesting to discover whether they actually exist scattered through many individuals in wild forms. In such circumstances chance would, in a proportion of cases which would be calculable were all the data known, produce individuals as extreme as any to be found under domestication. We have in fact to consider whether what has happened is not rather an alteration in the distribution of genes through a population than the introduction of new genes or of new mutations.

It is perhaps interesting in this connection to consider domesticated dogs. An examination of a considerable number of dog skulls and skeletons from prehistoric sites in Europe and the Near East suggests to me that very few, perhaps only one, wild species formed the foundation of the European breeds of dog. These now exhibit an extraordinary range of size and structure of every part but seem to be all perfectly fertile if artificial insemination is used. It is obvious that these variations in domesticated dogs include within their extremes all wild dogs, living or extinct, with very few exceptions, and those forms, like the amphicyonines, which are very remote from normal dogs. If this is true then either all the genes involved existed in the ancestral stock or there have been rapid and parallel mutations, perhaps accompanied by duplications of genes or by some method of their multiplication. If this is the actual condition, then it seems probable that these genes must have been unable to produce their present-day effects because of the unsuitability of the milieu in which they existed; that in fact they were controlled by modifiers.

If such conditions indeed obtained, then the process of evolution might largely depend on the removal of the inhibiting influence of these modifiers, perhaps because they came to be part of some hereditary mechanism concerned in the determination of other characters which were changing in the course of the animals' evolution. In these circumstances these assumed genes would become active and might, insomuch as they were present in the common ancestors of several stocks, be expected to lead to identical courses of evolutionary change in them. Natural selection would insure that only those changes which were advantageous would survive.

But there remain two features of the evolutionary story of amphibia which seem to be quite incapable of explanation on any adaptational basis. These are the steady dorsoventral flattening of the head and anterior part of the body and the progressive reduction of the cartilage bone in the skeleton. It is not easy to imagine in what manner the flattening of the head facilitates the transformation of an aquatic embolomerous labyrinthodont into an essentially terrestrial form. It is still more difficult to believe that still further flattening could render more easy the alteration of the terrestrial animal so involved into a secondarily but completely aquatic animal. In fact it seems in every way likely that this flattening of the amphibian head is brought about by internal factors not directly influenced in any way by the environment.

Even more interesting is the steady reduction of ossification which occurs in the history of very many amphibia. In the best-known case, among the labyrinthodonts proper, this reduction of the ossification exists only in cartilage or replacement bone, the membrane bones lying in or immediately below the skin not only showing no reduction but probably increasing in actual amount. That this is so is well illustrated by a comparison of the large Coal Measure embolomer *Eogyrinus* with a *Capitosaurus* of Lower Triassic age of about the same size, or even better with a *Cyclotosaurus* from the Upper Trias.

In *Eogyrinus* the neural cranium is completely bony, the only openings from the brain cavity being the necessary foramina for the cranial nerves and blood supply, and a pineal foramen. In cyclotosaurs the only bones left are the two exoccipitals, which must survive for obvious mechanical reasons. The dermal skull of *Eogyrinus*, though quite complete, is extremely thin, not much more than two millimeters thick in many places in a skull about thirty centimeters in length. In *Capitosaurus* and *Cyclotosaurus* the equally extensive skull is of considerably greater thickness, often twice as thick. In addition the area of the clavicles and interclavicle of the Triassic form is much greater than that of the Carboniferous animal, and they are again more massive.

Thus the phenomenon we are considering is peculiar to cartilage bones and does not depend on any shortage or unavailability of calcium, phosphorus, and other constituents of bone, or on a general failure of phosphatase production. It is not restricted to the skull; the vertebral column and limb bones of Stereospondyli are less well ossified than those of such Carboniferous Embolomeri as are known and than the same elements in a majority of the earlier Rachitomi.

Here again it seems impossible to imagine any environmental effect exercised, either directly or by the agency of natural selection. That this is the case seems to be made more certain by the fact that an identical reduction of cartilage ossification, often but not always unaccompanied by loss of membrane bone, occurs in many other groups of lower vertebrates. Professor Stensiö has shown that Devonian cephalaspids have in general less well-ossified neural crania than those from Downtonian rocks. It is evident that the Coal Measure *Megalichthys* has much less massive, though equally extensive, ossifications in its neural cranium than the Middle Devonian *Osteolepis*. The descendants of these fish, the coelacanths, are, as Stensiö has shown, thoroughly well ossified in Upper Devonian times, while the Cretaceous descendant, *Macropoma*, has small and fragile bones, and the still-living form apparently even less. The Middle Devonian *Dipterus* has a massive, bony neural cranium; the modern lung fishes, except perhaps in extreme old age, have no cartilage bones at all. The Upper Devonian and Carboniferous palaeoniscids have extremely well-ossified neural crania, those from the Trias very much less; the Lower Liassic sturgeon *Chondrosteus*, which is in effect their descendant, has a few small separated bones, while only the very large modern sturgeons have even traces. It is inconceivable that the steady loss of cartilage bones in each of these diverse groups, some of whose members lived in fresh water while others inhabited the sea, could have been due to external circumstances. It must depend on internal factors.

In this study I have tried to explain the methods of collecting data and of their direct interpretation which are used by paleontologists, and have endeavored to establish the intellectual respectability of these methods by showing that they have enabled us to make predictions which have subsequently been verified by new discoveries. By the use of these methods I have disentangled the history of certain groups of vertebrate animals, showing that a common feature of the histories, and indeed of most other phylogenies established by paleontologists, is that the course of evolutionary change in structure of some parts, or of the whole animal, is constant over a long period of time. The skeletons of most animals bear so many marks of the attachment of the muscles which surrounded them that it is often possible to restore the whole locomotory mechanism with considerable certainty and to ascertain the movements which are possible. If the various stages of an evolutionary series are so analyzed and compared with one another it will usually be found that the changes which take place are such as to fit the animals more perfectly, as regards their mechanical efficiency, for some special mode of life.

It is evident that such adaptive evolution is necessarily related to environmental conditions and that it can in principle be satisfactorily explained by natural selection. The mathematical investigations of Fisher and Haldane have shown that the effect of even a minute selective advantage is very great, even in what is from a geologist's standpoint a small number of generations. And studies of wild populations have shown the great abundance of recessive mutations existing in them.

Thus the only remaining postulate of the Darwinian theory whose validity remains to be established by observation is the existence of a selective death rate. The fact established by Haldane that the selective difference need only be extremely small, perhaps only one-tenth of one per cent, explains the difficulties that there have always been in making the necessary observations. But the fact which is now becoming clear, that the occurrence of mutations as phenotypes in wild populations is much rarer than would be expected from the proportion in which recessive mutant genes can be recovered from them, gives the required evidence in a satisfactory way. Thus many of the orthogenetic changes shown by studies of fossil vertebrates to occur can be satisfactorily accounted for by natural selection. Other changes are adaptive but are carried out not in relation to external conditions but in order to insure that the animals in which they take place remain workable; they meet mechanical or other needs which arise because of other changes which may, but need not, be themselves adaptive.

In the case of the amphibia these fundamental changes, on which the more obvious structural modifications are consequential, are very few in number, only two being well established, and are of a kind for which no adaptive explanation seems possible. They appear to depend on secular internal factors whose nature is entirely unknown.

All the types of change which we have so far considered cut across specific differences and seem to have an entirely different significance and mechanism. It is suggested that the origin of species is a phenomenon which has little to do with the main course of evolution, that it is dependent on accidental and localized occurrences.

APPENDIX

Since I delivered this series of lectures I have become aware of the possibility that many orthogenetic changes in the structure of animals may have been brought about as a result of some other secular change which may or may not be adaptive in nature. I believe the phenomenon to which I refer to be widely spread among animals, and it is most clearly exemplified by mammals. Consideration of the changes in proportion of the body during growth of any mammal suggests that they proceed regularly and are in principle capable of representation by some simple mathematical equation. In extreme cases of disproportionate growth, as for example in the horns of deer, the problem has been investigated by Huxley and others, but there is still no large accumulation of data dealing with the bodily proportions of more ordinary animals except man, though Mrs. Hacker (Miss Helga Pearson) has shown me some figures for the proportion between the humerus and radius in a very large series of the field mouse *Apodemus*. In the latter case it is obvious that the proportional change is of the utmost simplicity, figures for the length of the two bones plotted directly giving a straight line, which does not pass through the origin, and the scatter of the points being extremely small. We may therefore assume that there is a logistic of growth in all parts of the mammalian skeleton.

In one case, the proportion which the cranium bears to the face in horse skulls, Robb has shown that the changing proportion during growth of living horses can be expressed by a simple equation, and that fossil horses, from *Eohippus* onward, fall on the same curve. This would imply the persistence of whatever mechanism is responsible for the exact character of this growth throughout the whole history of the horse stock, and a further treatment of these figures by Simpson and Rau confirms this view. In measurements of this kind it is customary to compare one length with another, and it is evident that such a mode of procedure will lead to the simplest results. But in fact an individual length has very little meaning in the life of an animal, and I think it evident that the fundamental quality, to which all others are ultimately related, is size.

Size is a quality singularly difficult to define. An attempt at its interpretation may be made as follows. In order that it may live every animal must eat food and take in oxygen, and, in the case of mammals, the minimum amount of both food and oxygen necessary to maintain life can be determined, and is indeed the basal metabolism so widely examined by Benedict. The figure for the minimum necessary amount of oxygen is determined during sleep or complete quiescence. It represents that which is needed to keep the heart beating, the respiratory muscles working, the maintenance of a minimum tone in muscles, and similar fundamental needs. It has been shown by Benedict that with greater or less accuracy this minimum oxygen requirement under these conditions varies rather closely with the surface area of the individual experimental animal; and it has more recently been shown that a similar, or nearly similar, relationship holds for a number of invertebrate animals, e.g., the pupae of *Drosophila*. But other things being equal, the surface area is directly related to the weight, and it is really weight, suitably interpreted, which is the important matter.

It has, for example, become customary among agricultural physiologists to split up the food which a cow eats each day into two parts, the maintenance ration and the production ration. The maintenance ration representing that minimum amount which would be consumed by a cow (not lactating) for life without growth is weight. This figure is related to the live weight of the animal, varying essentially, like the basal metabolism, as the $\frac{2}{3}$ power of the weight; the best existing figure for cattle is actually the 0.73 power of the weight. The production ration is not in any way related to weight; it is expressed as containing so much protein and so much calorific value for each gallon of milk produced, and there are many small cows which produce far more milk than large ones. This mode of treating the food supply of milk cattle has proved of great value in practice, so much so that the rationing system for cattle food in Great Britain, during the war and since, was based upon it.

There is evidence that a similar procedure is applicable in the case of horses, where production of course means a distance run or a load hauled. It is evident that similar conditions exist in wild animals and hence, as the milk-producing function is never developed in them to the extent that it is in domestic cattle, and the main part of the production ration is expended in locomotion, it follows that the production ration will vary as the animal's weight, so that the total necessary food intake for any particular level of activity will be composed of one moiety related to a $\frac{2}{3}$ power of the weight, and the other to the weight itself, and in consequence a large mammal is thermodynamically more efficient than a small one of the same kind. Oxygen consumption follows food intake and is in turn related to the absorptive surface of the gut, the volume of the lung, and rate of respiratory movements. The volume of blood circulated per unit of time is clearly tied up with these other functions. Indeed it seems clear that weight is the most fundamental of all variable qualities of an animal; but weight is itself immensely complex. A fat ox may have 40 per cent of its live weight in fat, and these depot fats are not actively metabolizing; they do not use oxygen to any considerable extent, and their maintenance, as distinct from additions to them, requires very little food. They should clearly be excluded from the live weight in any judgment of size, but unfortunately, to a very large extent, they merely replace water in the body, and a considerable amount of fat may be laid down in the body of an ox without any equivalent increase in its live weight. What is wanted as a measure of size is an estimate of the metabolically active parts of the body, and perhaps the most suitable measure is the amount of nitrogen contained in it, the amount present in the hair and the hoofs being negligibly small. It is thus to weight so measured that we should refer other measurements made on an animal.

But unfortunately the total protein content of a large animal can be obtained only by immense labor, and is known only in cattle, sheep, pigs, and poultry. The sole practical procedure is to use live weight, but the live weight of an extinct animal cannot be determined directly. And though I believe that an investigation of the relationship between the live weights of horses, or any ungulates, and measurements of their metatarsals or some other limb bones might yield results of considerable interest, which would enable us to estimate the weight of a horse from a single bone with sufficient accuracy to be usable in dealing with an evolutionary series beginning with animals weighing perhaps ten pounds and ending with others two hundred times as heavy, such investigations have not yet been made. It seems, however, perfectly clear that such matters as the size of the antlers of deer, the horns of titanotheres, or the length of the lower jaw in mastodons (in addition to more ordinary variable structures, such as the length of limbs in relation to the vertebral column in a great many of the larger mammals) may well be related directly to the animal's weight and owe their proportions to an inherited mechanism which controls development. And judging from Robb's measurements of horses, and from the appearance of series of trilophodont and other mastodons, and similar series of other mammals, it appears that the observed conditions may well have been produced by the persistence through millions of generations of one unaltered mechanism. One of the difficulties will be to distinguish alterations in proportions dependent on such an unchanging mechanism from possibly very similar changes brought about by the direct action of natural selection.

BIBLIOGRAPHY

1. Amalitzski, B. 1921. Dvinosauridae. Petrograd.
2. Andrews, C. W. 1910. "Skeleton of Peloneustes," *Geol. Mag.*, 7.
3. ——— 1910. "Marine Reptiles of the Oxford Clay," Pt. 1, *Brit. Mus. Nat. Hist. Cat.*
4. Boonstra, L. 1934. "A Contribution to the Morphology of the Mammal-like Reptiles of the Suborder Therocephalia," *Ann. S. Af. Mus.*, *31*, Pt. 2, 215–267.
5. Broili, F. 1915. "Über *Capitosaurus arenaceus* Münster," *Centralbl. f. mineral. Geol. u. Pal.*, No. 19.
6. ——— 1930. "Über *Gemündina sturtzi*, Traquair," *Abhandl. d. Bayer. Akad. d. Wiss.*, N.S. No. 6.
7. ——— and Fischer, E. 1917. "*Trachelosaurus Fischeri*, nov. gen., nov. sp. Ein neuer Saurier aus dem Buntsandstein von Bernburg," *Jahrb. Königl. Preuss. Geol. Landes.*, 37, Pt. 1, sec. 3.
8. ——— and Schröder, J. 1934. "Beobachtungen an Wirbeltieren der Karrooformation": IV, Ein neuer Gorgonopside aus den unteren Beaufort-Schichten; V, Über *Chasmatosaurus van hoepeni* Haughton, *Sitz. d. Bayer. Akad. d. Wiss.*, pp. 209–264.
9. ——— 1935. "Beobachtungen an Wirbeltieren der Karrooformation": X, Über die Bezahnung von Trirachodon Seeley; XI, Über den Schädel von Cynidiognathus Haughton; XII, Über einige primitive Anomodontier-Schädel aus den unteren Beaufort-Schichten; XIII, Über die Skelettreste eines Gorgonopsiers aus den unteren Beaufort-Schichten; XIV, Ein neuer Vertreter der Gorgonopsiden-Gattung Aelurognathus, *Sitz. d. Bayer. Akad. d. Wiss.*, pp. 189–355.
10. ——— 1936. "Beobachtungen an Wirbeltieren der Karrooformation": XV, Ein Therocephalier aus den unteren Beaufort-Schichten; XVI, Beobachtungen am Schädel von Emydochampsa Broom; XVII, Ein neuer Anomodontier aus der Cistecephalus-Zone; XVIII, Über Cynodontier-Wirbel, *Sitz. d. Bayer. Akad. d. Wiss.*, pp. 1–76.
11. Broom, R. 1910. "A Comparison of the Permian Reptiles of North America with those of South Africa," *Bull. Amer. Mus. Nat. Hist.*, *28*, Art. 20, 197–234.
12. ——— 1912. "On a New Type of Cynodont from the Stormberg," *Ann. S. Af. Mus.*, 7, Pt. 5, No. 17, 334–336. "On Some Points in the Structure of the Dicynodont Skull," *ibid.*, No. 18, pp. 337–351.

13. ———— 1913. "Studies on the Permian Temnospondylous Stego-cephalians of North America," *Bull. Amer. Mus. Nat. Hist., 32,* Art. 38, 563–595.

14. ———— and Haughton, S. H. 1913. "On a New Species of Scym-nognathus (*S. tigriceps*)," *Ann. S. Af. Mus. 12,* No. 6.

15. ———— 1914. "Some Points in the Structure of the Diadectid Skull," *Bull. Amer. Mus. Nat. Hist., 33,* Art. 7, 109–114.

16. ———— 1929. "On Some Recent New Light on the Origin of Mam-mals," *Proc. Linn. Soc. N. S. Wales, 54,* Pt. 5, 688–694.

17. ———— 1930. "Notes on Some Labyrinthodonts in the Transvaal Museum," *Ann. Trans. Mus., 14,* Pt. 1.

18. ———— 1932. The Mammal-like Reptiles of South Africa and the Origin of Mammals. London, H. F. and G. Witherby.

19. ———— 1936. "On the Structure of the Skull in the Mammal-like Reptiles of the Suborder Therocephalia," *Phil. Trans., 226.*

20. Bulman, O. M. B. and Whittard, W. F. 1926. "On Branchiosaurus and Allied Genera (Amphibia)," *Proc. Zool. Soc. Lond.,* Pt. 2, pp. 533–579.

21. Bystrow, A. P. 1935. "Morphologische Untersuchungen der Deck-knochen des Schädels der Wirbeltiere": 1, "Mitteilung Schädel der Stegocephalen," *Acta Zool., 16,* 65–141.

22. ———— 1938. "Dvinosaurus als neotenische Form der Stegocepha-len," *ibid., 19,* 209–295.

23. ———— 1939. "Blutgefässystem der Labyrinthodonten," *ibid., 20,* 125–155.

24. ———— 1944. "*Kotlassia prima,* Amalitzky," *Bull. Geol. Soc. Amer., 55,* 379–416.

25. Bystrow, A. P., and Efremov, J. A. 1940. "*Benthosuchus sushkini* Efr., a Labyrinthodont from the Eotriassic of Sharzhenga River," *Acad. Sci. d. U.R.S.S. Travaux de l'Institut Paléont. 10,* 5–152.

26. Casserius, J. 1601. De vocis auditusque organis historia anatomica. Ferrara.

27. Case, E. C. 1911. Revision of the Amphibia and Pisces of the Permian of North America. Carnegie Inst. Washington.

28. ———— 1931. Description of a New Species of *Buettneria,* with a Discussion of the Brain Case. Museum of Palaeontology, Uni-versity of Michigan, 3, No. 11, 187–206.

29. ———— 1932. A Collection of Stegocephalians from Scurry County, Texas, *ibid., 4,* No. 1, 1–56.

30. ———— 1935. Description of a Collection of Associated Skeletons of *Trimerorhachis, ibid., 4,* No. 13, 227–274.

31. Credner, H. 1886. "Die Stegocephalen aus dem Rothliegenden des Plauen'schen Grundes bei Dresden," sec. VI, *Zeitsch. d. Deutsch. Geol. Gesells., 38,* 576–632.

32. —— 1890. "Die Stegocephalen und Saurier aus dem Rothliegenden des Plauen'schen Grundes bei Dresden": sec. IX, *ibid.*, *42*, 240–277.

33. —— 1893. "Die Stegocephalen und Saurier aus dem Rothliegenden des Plauen'schen Grundes bei Dresden": sec. X, *ibid.*, *45*, 639–704.

34. Damas, H. 1944. "Recherches sur le développement de *Lampetra fluviatilis*," *Archives de biol.*, 55.

35. Dempster, W. T. 1935. "The Braincase and Endocranial Cast of *Eryops megacephalus* (Cope)," *Jour. Comp. Neur.*, 62.

36. Efremov, J. A. 1929. "*Benthosaurus sushkini*, ein neuer Labyrinthodont der Permo-Triasischen Ablagerungen des Scharschenga-Flusses, Nord-Düna Gouvernement," *Bull. de l'Acad. des Sci. de l'URSS*, pp. 757–770.

37. —— 1933. "Über die Labyrinthodonten der U. de S.S.R.": II, "Permische Labyrinthodonten des früheren Gouvernements Wjatka," *Acad. Sci. d. U.R.S.S., Travaux de l'Institut Paléozoologique*, *2*, 117–164.

38. Efremov, J. A. 1939. "First Representative of Siberian Early Tetrapoda," *Comptes rendus de l'Acad. Sci. U.R.S.S.*, *23*, No. 1, 106–110.

39. —— 1940. "Die Mesen-Fauna der permischen Reptilien," *Neuen Jahrb. f. Mineral.*, *84*, 379–466.

40. —— 1940. "*Ulemosaurus Svijagensis*, Riab., ein Dinocephale aus den Ablagerungen des Perm der U.d.S.S.R.," *Nova Acta Leopoldina*, *9*, No. 59, 155–205.

41. —— 1946. "On the Subclass Batrachosauria—an Intermediary Group between Amphibians and Reptiles," *Bull. Akad. Sci. U.R.S.S.*, No. 6.

42. Efremov, J. A., and Zekkel, J. D. 1937. "Notes on the Permian Tetrapoda and the Localities of Their Remains," *Acad. Sci. d. U.R.S.S., Travaux de l'Institut Paléont.*, *8*, 1–63.

43. Eichwald, E. 1848. "Über die Saurier des Kupferführenden Zechsteins Russlands," *Bull. Soc. Impér. des Naturalistes de Moscou*, *21*.

44. Fraas, E. 1896. "Schwäbische Trias Saurier," *Mittheil. Kgl. Nat.-Cabinet zu Stuttgart*, no. 5.

45. ——1913. "Neue Labyrinthodonten aus der Schwäbischen Trias," *Palaeontographica*, *60*, No. 81, 275–294.

46. Goodrich, E. S. 1918. "On the Development of the Segments of the Head in Scyllium," *Quart. Jour. Micr. Sci.*, *63*

47. —— 1930. Studies on the Structure and Development of Vertebrates. London, Macmillan and Co.

48. Gregory, W. K. 1910. "The Orders of Mammals," *Bull. Amer. Mus. Nat. Hist.*, *27*.

49. ———— 1926. "The Skeleton of *Moschops Capensis*, Broom, a Dinocephalian Reptile from the Permian of South Africa," *ibid.*, *66*, Art. 3, 179–251.

50. ———— 1946. "Paréiasaurs versus Placodonts as Near Ancestors to the Turtles," *ibid.*, *86*, Art. 6, 281–326.

51. Haughton, S. H. 1918. "Investigations in South African Fossil Reptiles and Amphibia": Pt. 11, "Some New Carnivorous Therapsida," *Ann. S. Af. Mus.*, *12*, Pt. 6.

52. ———— 1924. "On Some Gorgonopsian Skulls in the Collection of the South African Museum," *ibid.*, *12*, Pt. 8, 499–517.

53. ———— 1925. "Investigations in South African Fossil Reptiles and Amphibia": Pt. 13, "On the Amphibia of the Karroo System," *ibid.*, *22*, Pt. 13.

54. ———— 1929. "On Some New Therapsid Genera," *ibid.*, *28*, Pt. 1, 55–78.

55. Heintz, A. 1931. "Revision of the structure of *Coccosteus decipiens*, Ag.," *Norsk Geol. Tidssk.*, *12*.

56. ———— 1932. "The Structure of Dinichthys: A Contribution to Our Knowledge of the Arthrodira," Bashford Dean Memorial Volume, Archaic Fishes, *Amer. Mus. Nat. Hist.*, Art. 4, pp. 111–241.

57. Herrick, C. J. 1899. "The Cranial and First Spinal Nerves of Menidia. A Contribution upon the Nerve Components of the Bony Fish," *Arch. Neurology and Psychopathology*, *2*, 1–289.

58. van Hoepen, E. C. N. 1915. "Stegocephalia of Senekal. O.F.S.," *Ann. Trans. Mus.*, *5*, No. 2, 125–149.

59. Holmgren, N. and Stensiö, E. A. 1936. "Kranium und Visceralskelett der Akranier, Cyclostomen und Fische," *Handb. d. vergleich. Anat.*, *4*.

60. von Huene, F. 1913. "The Skull Elements of the Permian Tetrapoda in the American Museum of Natural History, New York," *Bull. Amer. Mus. Nat. Hist.*, *32*, Art. 18, 315–386.

61. ———— 1922. "Beiträge zur Kenntnis der Organisation einiger Stegocephalen der Schwäbischen Trias," *Acta Zool.*, *3*, 395–460.

62. von Meyer, H. 1857. "Reptilien aus der Steinkohlen-Formation in Deutschland," *Palaeontographica*, *6*, 59–220, Pls. 9–23.

63. Neal, H. V., and Rand, H. W. 1936. Comparative Anatomy. Philadelphia, Blakiston Co.

64. Newton, E. T. 1893. "On Some New Reptiles from the Elgin Sandstones," *Phil. Trans. B*, *184*, 431–503.

65. Nilsson, T. 1937. "Ein Plagiosauride aus dem Rhät Schonens," *Lund Univers. Årsskrift.*, N. F. Avd. 2, *34*, No. 2. *Kungl. Fysiogr. Sällsk. Handl.*, N. F. *49*, No. 2.

66. ———— 1946. "A New Find of *Gerrothorax Rhaeticus* Nilsson, a Plagiosaurid from the Rhaetic of Scania," *Lund Univers.*

Årsskrift., N. F. Avd. 2, *42*, No. 10. *Kungl. Fysiogr. Sällsk. Handl.*, N. F. *57*, No. 10.

67. Nopcsa, F. 1928. "Palaeontological Notes on Reptiles," *Geol. Hungarica, 1*, 4–84, fasc. I.

68. Olson, E. C. 1937. "The Skull Structure of a New Anomodont," *Jour. Geol., 45*, No. 8, 851–858.

69. ———— 1941. "The Family Trematopsidae," *ibid., 49*, No. 2, 149–176.

70. ———— 1947. "The Family Diadectidae and Its Bearing on the Classification of Reptiles," *Fieldiana: Geology, 11*, No. 1. Chicago Nat. Hist. Mus., April 23, 1947.

71. Owen, R. 1881. *Fossil Reptilia of the Liassic Formations*, Palaeontographical Society, London.

72. Palmer, R. W. 1913. "Note on the Lower Jaw and Ear Ossicles of a Foetal Perameles," *Anat. Anz., 43*, 510–515.

73. Patten, W. 1901. "On the Structure and Classification of the Tremataspidae," *Mem. Acad. Imp. Sci. St. Petersb.*, Series 8, *13*, No. 5, 1–33.

74. Peyer, B. 1932. "Die Triasfauna der Tessiner Kalkalpen": III, "Placodontia"; IV, *"Ceresiosaurus calcagnii"*; V, *"Pachypleurosaurus edwardsii." Abhandl. d. Schweizerischen Pal. Gesells., 51–52*, 3–68.

75. ———— 1934. "Die Triasfauna der Tessiner Kalkalpen": VII, "Neubeschreibung der Saurier von Perledo," *ibid., 53, 54*, 3–140.

76. ———— 1939. "Die Triasfauna der Tessiner Kalkalpen": XIV, *"Paranothosaurus amsleri," ibid., 62*, 1–87.

77. Piveteau, J. 1934. "Les Poissons du Trias Inférieur. Contribution à l'étude des Actinopterygiens," *Ann. de paléontologie, 23*, 83–178.

78. ———— 1937. "Un Amphibien du Trias Inférieur. Essai sur l'origine et l'évolution des Amphibiens Anoures," *ibid., 26*, 135–177.

79. Price, L. I. 1935. "Notes on the Brain Case of Captorhinus," *Proc. Boston Soc. Nat. Hist., 40*, No. 7, 377–386.

80. ———— 1937. "Two New Cotylosaurs from the Permian of Texas," *Proc. New Engl. Zool. Club, 16*, 97–102.

81. Quenstedt, F. A. 1850. Die Mastodonsaurier im Grünen Keupersandsteine Württemberg's sind Batrachier. Tubingen, H. Laupp'schen Buchhandlung.

82. Riabinin, A. N. "A Labyrinthodont Stegocephalian *Wetlugasaurus angustifrons*, nov. gen., nov. sp., from the Lower Triassic of Vetlugaland in Northern Russia," *Yearbook of the Russian Paleontological Society, 8*, 49–76.

83. ———— 1927. *"Trematosuchus yakovlevi*, nov. sp. from the Lower Triassic Deposits in the Surroundings of Rybinsk," *Bull. du Comité Géol., 45*.

84. Romer, A. S. 1930. "The Pennsylvanian Tetrapods of Linton, Ohio," *Bull. Amer. Mus. Nat. Hist.*, 59, Art. 2, 77–147.

85. ———— 1933. Vertebrate Palaeontology. Chicago, University of Chicago Press.

86. ———— 1946. "The Primitive Reptile Limnoscelis Restudied," *Amer. Jour. Sci.*, 244, 149–188.

87. ———— 1946. "The Early Evolution of Fishes," *Quart. Rev. Biol.*, 21, No. 1, 33–69.

88. ———— 1947. "Review of the Labyrinthodonta," *Bull. Mus. Comp. Zool.*, Harvard College, 99, No. 1, 3–368.

89. Romer, A. S., and Witter, R. V. 1942. "Edops, a Primitive Rhachitomous Amphibian from the Texas Red-Beds," *Jour. Geol.*, 50, No. 80, 925–960.

90. Säve-Söderbergh, G. 1932. "Preliminary Note on Devonian Stegocephalians from East Greenland," *Meddelelser om Grønland*, 94, No. 7, 6–105.

91. ———— 1935. "On the Dermal Bones of the Head in Labyrinthodont Stegocephalians and Primitive Reptilia," *ibid.*, 98, No. 3, 5–211.

92. ———— 1936. "On the Morphology of Triassic Stegocephalians from Spitzbergen and the Interpretation of the Endocranium in the Labyrinthodontia," *Kungl. Svensk, Vetensk. Handl.*, 16, No. 1.

93. ———— 1937. "On the Dermal Skulls of Lyrocephalus, Aphaneramma and Benthosuchus, Labyrinthodonts from the Triassic of Spitzbergen and North Russia," *Bull. Geol. Inst. Upsala*, 27, 189–208.

94. Sollas, W. J. 1881. "On a New Species of Plesiosaurus (*P. conybeari*) from the Lower Lias of Charmouth," *Quart. Jour. Geol. Soc. Lond.*, 37, 440–481.

95. ———— 1916. "The Skull of Ichthyosaurus, Studied in Serial Sections," *Phil. Trans.* B, 208, 63–125.

96. ———— 1920. "On the Structure of Lysorophus, as Exposed by Serial Sections," *Phil. Trans.* B, 209, 481–527.

97. Sollas, Igerna, and Sollas, W. J. 1913. "A Study of the Skull of a Dicynodon by Means of Serial Sections," *Phil. Trans.* B, 204, 201–225.

98. ———— 1916. "On the Structure of the Dicynodont Skull," *ibid.* B, 207, 531–539.

99. Steen, M. C. 1930. "The British Museum Collection of Amphibia from the Middle Coal Measures of Linton, Ohio," *Proc. Zool. Soc. Lond.*, Pt. 4, No. 55, pp. 849–891.

100. ———— 1934. "The Amphibian fauna from the South Joggins, Nova Scotia," *ibid.*, Pt. 3, pp. 465–504.

101. ―― 1937. "On *Acanthostoma vorax*, Credner," *ibid.*, Ser. B, Pt. 3, pp. 491–499.

102. ―― 1938. "On the Fossil Amphibia from the Gas Coal of Nýřany and other Deposits in Czecho-Slovakia," *ibid.*, Ser. B, *108*, Pt. 2, 205–283.

103. Stensiö, E. A. 1927. "The Downtonian and Devonian Vertebrates of Spitzbergen": Pt. 1, "Cephalaspidae," *Det Norske Videnskaps-akademi i Oslo*.

104. ―― 1932. The Cephalaspids of Great Britain. Brit. Mus. Nat. Hist.

105. Strong, O. S. 1895. "The Cranial Nerves of the Amphibia, a Contribution to the Morphology of the Vertebrate Nervous System," *Jour. Morph.*, *10*, 1.

106. Sushkin, P. P. 1923. "Notes on *Dvinosaurus* (Stegocephalia, Rachitomi)," *Comptes rendus de l'Acad. Sci. Russie*, pp. 11–13.

107. ―― 1925. "On the Representatives of the Seymouriamorpha, Supposed Primitive Reptiles, from the Upper Permian of Russia, and on Their Phylogenetic Relations," *Occasional Papers, Boston Soc. Nat. Hist.*, *5*, 179–181.

108. ―― 1926. "Notes on the Pre-Jurassic Tetrapoda from Russia," *Pal. Hungarica*, *1*, 323–344.

109. ―― 1927. "On the Modifications of the Mandibular and Hyoid Arches and Their Relations to the Brain-case in the Early Tetrapoda," *Pal. Zeitsch.*, 8.

110. ―― 1936. "Notes on the Pre-Jurassic Tetrapoda from U.S.S.R.": III, "*Dvinosaurus Amalitzki*," *Acad. Sci. d. U.R.S.S., Travaux de l'Institut Paléozoologique*, 5, 43–91.

111. Swinton, W. E. 1927. "A New Species of Capitosaurus from the Trias of the Black Forest," *Ann. Mag. Nat. Hist.*, Ser. 9, *20*, 177–186.

112. ―― 1930. "Preliminary Account of a New Genus and Species of Plesiosaur," *ibid.*, Ser. 10, *6*, 206–209.

113. Watson, D. M. S. 1911. "The Skull of Diademodon, with Notes on Those of other Cynodonts," *ibid.*, Ser. 8, *8*, 293–330.

114. ―― 1912. "On Some Reptilian Lower Jaws," *ibid.*, Ser. 8, *10*, 573–587.

115. ―― 1913. "On Some Features of the Structure of the Therocephalian Skull," *ibid.*, Ser. 8, *11*, 65–79.

116. ―― 1913. "Further Notes on the Skull, Brain and Organs of Special Sense of Diademodon," *ibid.*, Ser. 8, *12*, 217–228.

117. ―― 1913. "*Batrachiderpeton lineatum*, a Coal-Measure Stegocephalian," *Proc. Zool. Soc. Lond.*, pp. 949–962, Pls., 1–2.

118. ―― 1914. "Notes on *Varanosaurus acutirostris*, Broili," *Ann. Mag. Nat. Hist.*, Ser. 8, *13*, 297–310.

119. ⸺ 1914. "*Procolophon trigoniceps*, a Cotylosaurian Reptile from South Africa," *Proc. Zool. Soc. Lond.*, pp. 735–747, Pls. 1–3.

120. ⸺ 1914. "Notes on Some Carnivorous Therapsids," *ibid.*, pp. 1021–1038.

121. ⸺ 1917. "A Sketch Classification of the Pre-Jurassic Tetrapod Vertebrates," *ibid.*, pp. 167–186.

122. ⸺ 1919. "On Seymouria, the Most Primitive Known Reptile," *ibid.*, pp. 267–301.

123. ⸺ 1919. "The Structure, Evolution and Origin of the Amphibia—the 'Orders' Rhachitomi and Stereospondyli," *Phil. Trans.* B, *209*, 1–73, Pls. 1 and 2.

124. ⸺ 1920. "On the Cynodontia," *Ann. Mag. Nat. Hist.*, Ser. 9, *6*, 506–524.

125. ⸺ 1921. "The Bases of Classification of the Theriodonta," *Proc. Zool. Soc. Lond.*, pp. 35–98.

126. ⸺ 1924. "The Elasmosaurid Shoulder Girdle and Fore Limb," *ibid.*, pp. 885–917.

127. ⸺ 1926. "The Evolution and Origin of the Amphibia," *Phil. Trans.* B, *214*, 189–257.

128. ⸺ 1929. "Carboniferous Amphibia of Scotland," *Pal. Hungarica, 1,* 221–252, Pls. 1–3.

129. ⸺ 1937. 'The Acanthodian Fishes," *Phil. Trans.* B, *228*, No. 549, 49–146, Pls. 5–14.

130. ⸺ 1940. "The Origin of Frogs," *Trans. Roy. Soc. Edinb., 60,* Pt. 1, 195–231.

131. ⸺ 1948. "Dicynodon and Its Allies," *Proc. Zool. Soc. Lond., 118,* 823–877.

132. Welles, S. P. 1943. "Elasmosaurid Plesiosaurs with Description of New Material from California and Colorado," *Mem. Univ. Calif., 13,* 125–254.

133. Westoll, T. S. 1942. "Ancestry of Captorhinomorph Reptiles," *Nature, 149,* 667.

134. ⸺ 1942. "Relationships of Some Primitive Tetrapods," *ibid., 150,* 121.

135. ⸺ 1943. "The Origin of the Primitive Tetrapod Limb," *Proc. Roy. Soc. Lond.* B, *131,* 373–393.

136. ⸺ 1943. "The Hyomandibular of *Eusthenopteron* and the Tetrapod Middle Ear," *ibid.*, B, *131,* 393–414.

137. ⸺ 1943. "The Origin of the Tetrapods," *Biol. Rev., 18,* 78–98.

138. ⸺ 1944. "New Light on the Mammalian Ear Ossicles," *Nature, 154,* 770.

139. ⸺ 1945. "The Mammalian Middle Ear," *Nature, 155,* 114.

140. White, E. I. 1935. "The Ostracoderm *Pteraspis* Kner, and the Relationships of the Agnathous Vertebrates," *Phil. Trans.* B, *225,* No. 527, 381–457.

141. White, T. E. 1939. "Osteology of *Seymouria baylorensis,* Broili," *Bull. Mus. Comp. Zool.,* Harvard College, 85, No. 5, 325–409.

142. ——— 1940. "Holotype of *Plesiosaurus longirostris,* Blake, and Classification of the Plesiosaurs," *Jour. Pal., 14,* No. 5, 451–467.

143. Whittard, W. F. 1928. "On the Structure of the Palate and Mandible of *Archegosaurus decheni,* Goldfuss," *Ann. Mag. Nat. Hist.,* Ser. 70, *1,* 255–264.

144. ——— 1930. "The Structure of *Branchiosaurus flagrifer,* sp. n., and Further Notes on *Branchiosaurus amblystoma,* Credner," *Ann. Mag. Nat. Hist.,* Ser. 10, 5, 500–512.

145. Williston, S. W. 1906. "North American Plesiosaurs. Elasmosaurus, Cimoliasaurus and Polycotylus," *Amer. Jour. Sci., 21,* Art. 16, 221–236.

146. ——— 1907. "The Skull of Brachauchenius, with Observations on the Relationships of the Plesiosaurs," *Proc. U. S. Nat. Mus., 32,* 477–489.

147. ——— 1910. "New Permian Reptiles: Rhachitomous Vertebrae," *Jour. Geol., 18,* No. 7, 585–600.

148. ——— 1911. "Restoration of *Seymouria baylorensis* Broili, an American Cotylosaur," *ibid., 19,* No. 3, 232–237.

149. ——— 1911. "American Permian Vertebrates. University of Chicago Press.

150. ——— 1913. *Ostodolepis brevispinatus,* a New Reptile from the Permian of Texas," *Jour. Geol., 21,* No. 4, 363–366.

151. ——— 1914. "The Osteology of Some American Permian Vertebrates," *ibid., 22,* No. 4, 364–419.

152. ——— 1915. "Trimerorhachis, a Permian Temnospondyl Amphibian," *ibid., 23,* No. 3, 246–255.

153. ——— 1915. "A New Genus and Species of American Theromorpha. *Mycterosaurus longiceps," ibid., 23,* No. 6, 554–559.

154. ——— 1916. "Synopsis of the American Permocarboniferous Tetrapoda," *Contributions from Walker Museum,* University of Chicago Press, *1,* No. 9, 193–236.

155. ——— 1918. "The Osteology of Some American Permian Vertebrates": III, *ibid., 2,* No. 4, 87–112.

156. ——— 1925. The Osteology of the Reptiles. Harvard University Press.

INDEX

Acanthodes, gill region, 17
Acanthodian, branchial region, 15
Acanthodii, 8–10
Acheloma casei, brachyopid, 52
Actinopterygii, evolution of, 105
Adelospondyli, 91; vertebrae, 38
Amphibia, eggs laid in water, 105–106; evolution of, 136; flattening of, 195; from Linton, 89; life history, 106; nature of evolutionary history, 105; reduced ossification, 195; variation in skull shape, 35; vertebrae, 49–52; vertebral column, 36–38
Anatomy of cynodont head, 156
Animals, size basis of measurement, 200; structure susceptible to mechanical analysis, 182; weight the true measure of size, 201
Apodemus, proportions, 200
Archegosaurus, 78; gill rakers, 100; growth of skull, 36
Atthey, Thomas, 81

Batrachiderpeton, 90
Batrachosuchus, 53–55
Benedict, basal metabolism, 201
Brachyopids, history, 53; palates, 54; persistent characters, 62
Branchiosaurs, gills, 99; metamorphosis, 99
Brithopus, derived from sphenacodont ancestors, 135
Bulls, not nicking with cows, 196
Bystrow, on *Dwinasaurus*, 104

Capitosaurs, evolutionary series, 41–42; evolution of lower jaw, 67–69, of occiput, 47, of palate, 46; flattening of skull, 71; mechanical effects of evolution, 66–77; occipital condyle, 47; persistent qualities, 60–61; skull struc-

ture, 43–45; widened occiput, 69–70
Capitosaurus, lower jaw, 69
Captorhinus, stapes, 134; tympanic membrane? 133
Cartilage bone in Actinopterygii, 198; cephalaspids, 198; coelacanths, 198; Dipnoi, 198; labyrinthodonts, 197; osteolepids, 198
Cartilage, delayed ossification of, 195
Casserius, on ear ossicles, 137
Cephalaspis, feeding habits, 28; mechanism of breathing, 29; mouth, 27; prehyoidean gills, 30; segmentation of head, 13; structure, 24–27
Ceraterpeton, 89
Changes of structure consequential to others, 194
Chromosomes, 186–187
Ciliary feeding, 22
Climatius, evidence as to structure, 14–15
Coal Measures, mode of formation, 33
Coecilia, no fossil history, 97
Colosteus, related to Ichthyostega? 103
Compensation, in evolution, 59
Constant direction of evolutionary change, 194
Continental deposits, 33
Cotylosaurian vertebrae, 112
Cotylosaurs, origin of, 126; polyphyletic? 126; primitive reptiles, 115
Cyclotosaurus, habits, 65; lateral line, 66; lower jaw, 68; mechanics of skull, 73–74
Cynodont, anatomy of head, 156; lower jaw, 153; mechanism of

213

Cynodont (*continued*)
jaw, 157–159; muscular cheek and skin, 125; secondary palate, 124; sense of hearing in, 125; skeleton, 123; turbinals and warm bloodedness, 125

Darwinian evolution, 190; postulates, 193
Dendrerpeton, intermediate between Embolomeri and Rachitomi, 87
Deuterosaurus, derived from sphenacodont, 135
Diadectes, structure of ear region, 130–131; structure of skull roof, 130
Diademodon, external auditory meatus of, 164
Differential fertility, 192
Digastric muscle, in cynodonts, 164
Dimetrodon, elongated neural spines, 121; origin of dentition and palate, 121
Diplocaulus, 90–91
Drosophila, recessive mutations in wild, 192; results of crosses calculable, 185
Dwinasaurus, 53–55; description, 77; perennibranchiate, 101–102

Ear of cyclostomes, function, 138; of lizard, 140–141; of mammal, 141–143
Edops, antecedent to *Eryops,* 101; related to *Eryops,* 52
Elasmosaurus, characters of, 167–168; history, 169–170
Elasmosaur, shoulder girdle, 174
Elphistostege skull, 103
Embolomeri, shoulder girdle, 86–87; two divisions of, 101
Eogyrinus, locomotion, 85; skeleton and habits, 84–85
Epipterygoid, of mammal-like reptiles, 153–154
Erpetosuchus, related to Ichthyostega? 103
Eryops, habits, 65; lower jaw, 69;

mechanism of skull, 72; predicted ancestor, 79–80
Evolution, constant direction of, 194, 198; dependent on new genes? 196; nonadaptive, 194; parallel, 105; restricted range of possibilities, 195
Evolutionary trend and transformation of function, 136
External auditory meatus of *Diademodon,* 164
External gills, 97–99
Extracolumella, absent in mammal-like reptiles, 162

Fish-eating labyrinthodonts, 55–58; artificial series, 62–63
Fisher, R. A., on *Gryllotetrix,* 192
Fossils, early recognition of, 1; establish age of beds, 32
Frogs, 97

Genealogical trees, method of making, 2, 3
Genes, position effect, 195; with additive effect, 196
Geographical races, 186
Gerrothorax, Rhaetic brachyopid, 52
Gills, introduction of, 23
Gorgonopsids, dentition, 123; skull and its mode of origin, 148–152; tympanic cavity of, 152–153
Growth, mechanism determining, 200

Haldane, J. B. S., effect of selective advantage, 192
Hearing in amphibia, 138; in animals without a tympanic membrane, 118; in mammals, 136
Herring investigations in North Sea, 191
Homology, significance of, 3
Horses, skull proportions of, 200
Huxley, J. S., on growth, 200
Huxley, T. H., on Carboniferous amphibia, 81
Hyoid in *Diadectes,* 133; loxom-

mids, 132; seymouriamorphs, 133

Ichthyosaur evolution, 179
Ichthyostegids, 92–97; evolutionary history, 103; skull pattern, 97

Jagorina gill region, 18–19

Kotlassia and *Seymouria*, 126

Labyrinthodonts, cartilage bone in, 197; characteristic ornament, 33–34; classification, 38–39; decreased ossification in later, 76–77; epipterygoid, 74–76; first found Carboniferous, 80; habits, 64; membrane bone in, 76–77; neoteny in, 77; parallel evolution in, 59; skull, mechanics of, 72–74; structure of Carboniferous, 82–84; vertebrae, 36–37
Lamarckian evolution, 190
Lanthanosuchus, temporal fossa, 128; seymouriamorph, 128
Lepospondyli, 89; vertebrae, 38
Limnoscelis, ear of, 134; systematic position of, 129
Lonchorhynchus, Säve-Söderbergh on, 77

Mammal, daily food consumption, 201; maintenance ration, 201; origin of jaw articulation, 158–160; production ration, 201; thermodynamical efficiency of, 202; weight, how built up, 202; weight estimated from measurements, 202
Mammal-like reptiles, skull, 118; temporal fossa, 119
Marsupial, skull of pouch young compared with cynodont, 155
Mastodons, size and structure, 202
Mechanical efficiency, increase of, 199
Membranous labyrinth in cynodonts, 160
Mesacanthus, branchial region, 16

Metoposaurus, delayed ossification of, 76
Milk yield, how determined, 195–196
Morphology, nature of, 3
Mouth, comparison with gill slit, 14
Musculus depressor mandibuli, disappearance in late mammal-like reptiles, 163

Neoteny in labyrinthodonts, 77
Nothosaurs, 169, 170

Orthogenetic evolution, 184
Osteolepids, as ancestors of amphibia, 95

Paleontology, Darwin on, 1; independence of, 1; respectability of methods, 198
Paleontologist, limitations of, 188–189
Palato-quadrate cartilage, evolution of in mammal-like reptiles, 143
Parallel evolution, 183, 184, 195
Pariasaurs, 131; origin of, 132
Pelycosaurs, absence of tympanic membrane, 133, 161
Perennibranchiata, 97
Permian, reptiles of basal, 115
Phyllospondyli, branchiosaurs, 88; derivatives of Embolomeri, 89
Phyllospondyl, vertebra, 38
Phylogenetic trees, how made, 31
Piveteau, J., *Protobatrachus*, 103, 104
Plagiosaurus, 53–55
Platyceps, larval brachyopid, 100; external gills, 101
Platyops, 77
Plesiosaurs, age and distribution, 165–166; long-headed, 176–178; shoulder girdle, 171–173
"*Plesiosaurus*" conybeari, 169
Postulates for a Darwinian explanation, 193
Procolophon, ear region, 132
Procolophonids, 131

Procynosuchus, 135

Protochordates, feeding methods, 22

Pteraspis, gill structure, 29

Rachitomous vertebrae, origin of, 51

Recessive mutation in wild *Drosophila,* 192

Reproductive isolation of species, 185–186

Reptiles, early history of, 115; early separation of two branches, 116; eggs, 106; life history, 106

Rhamphodopsis, gill region, 18

Rhinesuchus, 47

Rhopalodon (Brithopus), 135

Robb, proportions of horse skulls, 200

Romer, A. S., on Carboniferous phyllospondyls, 103; on trematosaurs, 78

Russian mammal-like reptiles, 135

Sediments, superposition as evidence of age, 31

Segmentation of head, Balfour on, 13, 19–22; somites, 11

Segmentation of trunk muscles, 23; of vertebrate head, 4–7

Selective advantage, effectiveness of small, 199

Selective death rate in nature, 191

Seymouria, 107–113; lateral line of, 114; life history, 115; lower jaw, 110; palate, 109; pectoral girdle, 113; skull, 108; tympanic cavity, 115–116; vertebral column, 110–111

Seymouriamorphs, group, 114; reptilian qualities of, 129

Shoulder girdle of plesiosaurs, 171–173

Size of animals, a fundamental quality, 200; how defined, 200, 201

Skull of mammal and reptile compared, 139

Smith, William, geological maps, 32

Species, criteria of, 188; nature of, 185

Specific differences, imposed on general evolution? 199

Stapes, derived from hyomandibular, 138; of labyrinthodonts, 139

Stratified rocks, how laid down, 31; order of succession, 32

Sushkin, on *Seymouria,* 114

Theriodonts, classification of, 135

Therocephalia, in Russia? 135

Time scale, stratigraphical basis, 31

Trachelosaurus, 169

Trends, evolutionary, 64; projected backward, 79

Tritylodon, is an ictidosaur, 135

Tympanic membrane. 138

Urals, copper-bearing sandstones, 135; mammal-like reptiles from, 122

Urodela, structure, 97

Varanops, general structure, 119–121

Wetlugasaurus, 52

Zoarces, Schmidts' work on, 191

SILLIMAN MEMORIAL LECTURES
Published by Yale University Press

Electricity and Matter. By Joseph John Thomson. [OUT OF PRINT.]

The Integrative Action of the Nervous System. By Charles S. Sherrington.

Experimental and Theoretical Applications of Thermodynamics to Chemistry. By Walter Nernst. [OUT OF PRINT.]

Radioactive Transformations. By Ernest Rutherford. [OUT OF PRINT.]

Theories of Solutions. By Svante Arrhenius. [OUT OF PRINT.]

Irritability. By Max Verworn. [OUT OF PRINT.]

Stellar Motions. By William Wallace Campbell.

Problems of Genetics. By William Bateson. [OUT OF PRINT.]

The Problem of Volcanism. By Joseph Paxson Iddings. [OUT OF PRINT.]

Problems of American Geology. Dana Commemorative Lectures. [OUT OF PRINT.]

Organism and Environment as Illustrated by the Physiology of Breathing. By J. S. Haldane. [OUT OF PRINT.]

A Century of Science in America. By Edward Salisbury Dana and others. [OUT OF PRINT.]

The Intestinal Flora. By Leo F. Rettger and Harry A. Chaplin. [OUT OF PRINT.]

The Evolution of Modern Medicine. By Sir William Osler.

Respiration. By J. S. Haldane. [OUT OF PRINT.]

After Life in Roman Paganism. By Franz Cumont. [OUT OF PRINT.]

The Anatomy and Physiology of Capillaries. By August Krogh. [REVISED EDITION.]

Lectures on Cauchy's Problem in Linear Partial Differential Equations. By Jacques Hadamard. [OUT OF PRINT.]

The Theory of the Gene. By Thomas Hunt Morgan. [OUT OF PRINT.]

The Anatomy of Science. By Gilbert N. Lewis.

Blood: A Study in General Physiology. By Lawrence J. Henderson.

On the Mechanism of Oxidation. By Heinrich Wieland.

Molecular Hydrogen and Its Spectrum. By Owen Willans Richardson. [OUT OF PRINT.]

The Changing World of the Ice Age. By Reginald Aldworth Daly. [OUT OF PRINT.]

The Realm of the Nebulae. By Edwin Hubble. [OUT OF PRINT.]

Embryonic Development and Induction. By Hans Spemann. [OUT OF PRINT.]

Protein Metabolism in the Plant. By Albert Charles Chibnall. [OUT OF PRINT.]

The Material Basis of Evolution. By Richard Goldschmidt. [OUT OF PRINT.]

Centennial of the Sheffield Scientific School. Edited by George A. Baitsell.

Elementary Particles. By Enrico Fermi.

Paleontology and Modern Biology. By D. M. S. Watson.